THE
CONSTANT
ECONOMY

Born in 1975, Zac Goldsmith joined the *Ecologist* magazine in 1997 and became its editor. In 2004 he received Mikhail Gorbachev's Global Green Award for 'International Environmental Leadership'. He is now the Conservative Member of Parliament for Richmond Park and North Kingston in London. For more information go to: www.theconstanteconomy.com.

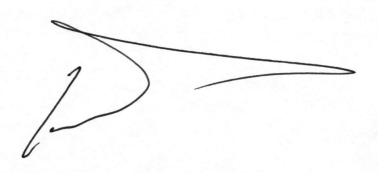

THE CONSTANT ECONOMY

How to Create A Stable Society

Zac Goldsmith

Atlantic Books
London

First published in Great Britain in hardback and airport and export trade paperback in 2009 by Atlantic Books, an imprint of Grove Atlantic Ltd.

This paperback edition published in Great Britain in 2010
by Atlantic Books.

1 3 5 7 9 10 8 6 4 2

A CIP catalogue record for this book is available from the British Library.

ISBN: 978 1 84887 095 6

Printed in Great Britain

Atlantic Books
An imprint of Atlantic Books Ltd.
Ormond House
26–27 Boswell Street
London
WC1N 3JZ

www.atlantic-books.co.uk

For Uma, Thyra and James

Contents

Preface

The first British election fought on environmental issues was hardly an extravagant affair. It didn't capture the imagination of the public or the world media, nor was it fought by a recognizable political party. Yet in terms of the development and progress of green politics, it was a key event – even if the unlikely battleground was Suffolk.

In 1974, the global environment was, at best, a marginal concern. And in one of the most keenly contested general elections of the twentieth century, the second in the same year, it was going to take something startling for green issues to be taken seriously. My uncle Teddy – founder of the *Ecologist* magazine and member of 'People', which eventually became the Green Party – helped draw up a stark manifesto based around his magazine's *Blueprint for Survival*. It was obvious to him that he would need something more than mere argument, or even the snappy line 'No deserts in Suffolk. Vote Goldsmith' to attract people's attention. He needed a camel.

Teddy managed to find one, and it provided much-needed colour in an otherwise greyish political climate. As a stunt, however, it failed to pull in the voters, and Teddy – to no one's surprise – lost his deposit in style. Not only that, but a paper-waving official accused him of animal cruelty, citing the effects on the camel of breathing in car fumes. 'That's exactly my point,' Teddy declared. 'Imagine what it's doing to us!'

Thirty-five years later and there has been a seismic shift. Green concerns have moved from the fringes of political debate into the mainstream of government. But despite this progress, there remains an almighty gulf between what is said and what is done. Tony Blair, for instance, described climate change as 'the greatest long-term threat to our planet'. 'Inaction', he said, would be 'literally disastrous'. In charge for a decade, Blair had presided over a country that became neither less polluting, nor more prepared for environmental change.

There are nevertheless patches of good news from the world of politics – many of which are mentioned in the course of this book. The trouble is that most of them address only one, albeit immense symptom of the environmental crisis: climate change. They do little to address the fact that we are rapidly shifting from an era of abundance towards one of scarcity – a situation caused by a combination of massive population growth, an insatiable human appetite for consumption and an ever-shrinking resource base.

This might seem like a nightmarish vision of the future, but it is, in fact, a mathematical certainty. We cannot continue to consume the world's resources at the rate we are, without expecting them to run out at some point. But that very basic truth has almost no bearing on policy decisions. Governments shy away from tackling the issue, terrified of antagonizing voters with unpopular policies. The underlying assumption is that there is a straight choice between economy and ecology – and, ultimately, the economy always wins. But it's a false choice.

The recession has cost many people their jobs, their savings and even their homes. In such times, concern for the environment necessarily slips down the agenda. But the right environmental solutions would help, not hinder people struggling to cope. And when we emerge, as we know we will, we can do so with an economy that is environmentally literate, where green choices that are currently available only to the wealthy become available to all.

Now is the time to decide what sort of economy we want to develop from the ashes of this recession. Instead of struggling to recreate the conditions that delivered it, we can choose to stimulate the development of a cleaner, greener and much less wasteful economy. We can build something new, something that will regenerate our stagnant economies, and which, unlike the growth model that has dominated for decades, can actually last. We ignored economists' warnings that we were living beyond our financial means. We cannot continue to ignore scientific warnings that we have been delving into nature's capital for too long. As one US conservationist has cautioned, 'Mother Nature doesn't do bailouts.'

Critics of the environmental agenda claim the cost of a green economy would be hundreds of billions, if not trillions of pounds. But they confuse cost with investment. For example, if I invest one hundred units in improving the energy efficiency of my local school, and save twenty units each year thereafter as a result, that represents a hugely rewarding investment opportunity. And the shift doesn't require 'new' money.

There should be no need for net tax increases to pay for our indulgence in things green. It simply requires bullish signals from government. If a proper cost is attached to pollution and waste, businesses will minimize both. And if the funds raised from taxing these activities are used to incentivize the opposite, we will see a dramatic shift in the movement of money towards the kinds of investments and activities that we need. With the right encouragements, whole sectors could flip. UK pension funds, for instance, control about £860 billion. Imagine the impact if they chose to invest it in the new green economy?

But it's not just the way that we invest that needs to be

3

addressed, it's the way that we look at costs. As a young child, I would pilfer ice cream from my mother's kitchen and sell it at knock-down rates to passers-by on the streets outside my home. The cash tin overflowed, and I was delighted. With zero capital costs on the balance sheet, I'd turned, relatively speaking, an enormous profit. It was only when my clandestine enterprise was uncovered that I was forced to confront the subsidies my parents had unwittingly provided and discovered that I had, in fact, made a substantial loss. If polluting industries had to pay for the clean-up, they would see a similar effect on their balance sheets. . .

Our politicians need to understand that reconciling the market with the environment is our defining challenge. And that it is possible. By shifting taxes, removing perverse subsidies and creating clear signals, this will happen naturally. Opportunities will spring up, jobs will be created and we will enjoy the emergence of a truly constant economy. By and large, Westminster knows this, so why do they remain so reticent?

One of the factors that most inhibits politicians is a media that remains hostile to green issues. How else to explain the baffling experience of opening the *Sun* newspaper one day to find a photograph of myself next to the image of a pink vibrator and under the headline 'Goldsmith Wants to Ban Dildos' (because sex toys are energy inefficient)? No less than the *Sun*'s main political editor angrily demanded that my ideas be 'dropped like a stone'. Of course he knew I'd never said anything of the sort. Indeed until the story appeared I had never spoken, let alone written, about sex aids. But for the *Sun* – and many other media outlets – green solutions are bad. They have to be bad, and they have to be stopped.

Of all the things to worry about, being accused of wanting to ban energy-inefficient vibrators isn't top of my list.

However, what is worrying is the reluctance of some of the most powerful media outlets to look seriously at green issues. For vote-dependent politicians, the treatment of environmental policy by the papers is reason enough to pause. But while it's easy to point the finger at the papers, I believe greens themselves have to shoulder some of the blame.

During their days in the wilderness, greens had to talk up the impending ecological crisis. They felt they had to shock people and went out of their way to scare people into action. While the world was looking away, uninterested in their prognosis, there was little else they could do. But as the world finally began to take heed, green voices factionalized, splitting into two quite different camps – both of them, to my mind, wrong.

The 'lighter' greens took the softer, more culturally agreeable route to green consumerism. If everyone switched to energy-efficient light bulbs, and drove better cars and bought better food, they said, the world would be saved. It's an attractive philosophy, but one that is ultimately flawed. Yes, the more people use green goods, the better for the planet, but it would require the vast majority of the world's people to change their lifestyles for the planet to feel a measurable impact. For any number of reasons this will never happen. Green choices need to be the norm, not the expensive gestures of a few who are committed or wealthy enough to make them. For all their good intentions, in trying to promote their impossible world consensus, 'lighter' greens are simply letting politicians off the hook.

'Darker' greens took a different path. Years of tracking the brutal consequences of market failure have nourished in them an understandable contempt for the market itself. Like our foot-dragging politicians and reluctant media commentators, they also believe, wrongly, that we are faced with a choice

5

between the economy and ecology. The only difference is that they favour the latter.

They have seen the market's transformative power, but they cannot imagine it being used for renewal, and they long instead for its replacement. But it is an illogical approach. Just as uncontrolled cell growth defines cancer, indiscriminate economic growth devastates the planet. The 'market' is no more to blame for environmental destruction than healthy cells are to blame for cancer. The problem is our failure to write the rules. Given that the market isn't going anywhere, that is what we must do.

But 'darker' greens have grown used to being at the radical fringe, and as mainstream society has crept closer, they have drifted away. Their alarm is extreme, their pessimism infectious and their disenchantment a dampener on the enthusiasm of ordinary people. When they identify solutions, they identify the hardest, most punitive solutions, and when they describe the challenge, it is invariably insurmountable. As a consequence, many people feel impotent and fatalistic in the face of the environmental challenge.

Both strategically and factually this is extremist dogma, and it provides environmental naysayers with the straw man they need to discredit the environmental agenda. We know we consume way beyond our means – if the experts are right, roughly three times beyond our means – but that doesn't mean we must live lives that are three times poorer. It means we should demand food that has travelled shorter distances, less packaging for goods, and products that will actually last. It means using taxes to protect natural capital, like forests and fisheries, so that we can continue enjoying the interest.

The temptation is to believe, as our politicians and much of the media believe, that if there's no pain, there can be no gain. For years, I certainly thought so – which might go some way to explaining why the *Ecologist*, which I edited for nearly a decade, became for a while, perhaps the world's gloomiest magazine. But my outlook was changed when I was asked by the British Conservative Party to help oversee a review of environmental policy with John Gummer MP.

Our job was to look for solutions both at home and abroad – to identify successful schemes and bright ideas. We discovered that almost everything that needs doing is already being done somewhere in the world. Looking at the portfolio of ideas we'd found, we saw that if we took the best of today in every sector and made it the norm tomorrow, we'd be halfway or further to our goal. I was struck by the simplicity of the solutions. Solutions that would actually help people cope with hard times, not add to their difficulties. Time and again, we'd stumble across something so obvious, and so effective, that we'd wonder why on earth it hadn't already been adopted.

Many of those ideas appear in this book. It is in no way exhaustive, but collectively these solutions offer a programme of action that could set us on course towards a healthy, constant economy; one that recognizes the inescapable link between nature and the economy, one that knows limits and can last.

Two hundred years ago, Edmund Burke, the father of conservative philosophy, said 'Society. . . becomes a partnership not only between those who are living, but between those who are living, those who are dead, and those who are to be born. Each contract of each particular state is but a clause in the great primeval contract of eternal society.' It's difficult to imagine a more sensible approach, nor one further removed from that of our current political leaders.

British politicians, and the British people, have it within them to rise to this challenge. They have done it before. In 1939, a whole generation fought what seemed like an impossible battle – and won. After victory, in 1945, that generation joined with an unprecedented, government-led mission to build a pioneering welfare state, which lifted millions out of poverty and revolutionized the lives of ordinary people. The disaster of war spurred us on to create new priorities, and build a better country. Today, the impending ecological disaster gives us the chance to rise to that challenge again.

The country needs leadership from its politicians, but they will not provide it unless we – the electorate – send them a clear message. For doing the right thing, they will be rewarded. For doing the wrong thing, they will be sacked and history will be harsh in its judgement. It is up to them to act, but we must make them act.

This book is not a self-help guide for improving individual lifestyles. It is a political programme: a tool for voters, and a challenge to the political classes; a gauntlet thrown down at their feet. They know what is wrong, and they know they must act. Here is what they should do. Here is the programme. If they don't agree with it, they must provide another way of achieving these goals – and then they must put it into practice. What they can no longer do is avoid these issues. Future generations will not forgive them.

Introduction

The Case for Change

*Whether we and our politicians know it or not,
Nature is party to all our deals and decisions, and
she has more votes, a longer memory, and a sterner
sense of justice than we do.* Wendell Berry

There can be no doubt that the natural world, on which we depend for each and every one of our needs, is in very serious trouble. We can argue about aspects of climate science, and we can quibble with some of the predictions: after all, there is no computer model in the world that can truly take into account the full complexity of ecological systems. But the looming environmental crisis is a basic observation, not a theory.

In 2005, the UN conducted a wide-scale audit of the planet's health. Its conclusions were stark. 'Over the past fifty years,' it reported, 'humans have changed ecosystems more rapidly and extensively than in any comparable period of time in human history, largely to meet rapidly growing demands for food, fresh water, timber, fibre and fuel. This has resulted in a substantial and largely irreversible loss in the diversity of life on Earth.'

Its findings make for sobering reading. Between 1970 and 2003 the population of land species declined by nearly a third, while populations of tropical species declined by more than

9

half. In the same period, humanity destroyed almost half the planet's original forests.

We are altering the very systems upon which we depend. Without coral reefs and mangroves to act as 'fish nurseries', fish stocks simply collapse. Without certain species of bee or wasp, many plants cannot be pollinated and will not grow. Without rainforests, the planet loses not only thousands of as-yet-undiscovered species, but also a 'carbon sink' that helps slow climate change.

At the root of all this is some simple mathematics. The human population is growing, along with our hunger for resources – but the Earth itself isn't. It's an uncomfortable fact, but it is nevertheless inescapable. Oil will eventually run out, and what remains is in the hands of countries we can't always rely on. The world's great breadbaskets are shrinking at an alarming rate, and water shortages now affect more than a hundred countries. All this, and there remains the biggest environmental challenge of all – climate change.

What was once a marginal scientific debate has become the framing argument for all our discussions about the future. If even the most conservative predictions are accurate, the effects will be serious – just how serious depends on how fast we act now to stave off the worst of its effects. When an organisation like Red Cross International warns that aid will not be able to keep pace with the impacts of climate change, we should be concerned. Still more so when major financial institutions issue similar warnings.

According to German re-insurers Munich Re, the economic losses from natural disasters increased eightfold from the sixties to the nineties. About 80 per cent of this resulted from extreme weather-related events. The company now predicts that by 2065, the cost of damages will outstrip global assets. Insurers of the United Nations Environment Programme

(UNEP), meanwhile, believe worldwide losses linked specifically to climate change will reach a yearly £184 billion in fifty years' time. It is the insurance industry's function to put a price on danger. Their warnings cannot simply be brushed aside.

In his report to the British government in 2006, the former World Bank economist Nicholas Stern described climate change's potential for major economic disruption and social chaos. The cost of delaying action, he said, is far greater than we can accommodate, and the longer we delay, the higher those costs will be.

But while climate change is the biggest problem we face, it is essentially a symptom of our dysfunctional relationship with the planet. Even if climate change were not happening, we would still need to address the fact that our water consumption globally is growing at twice the rate of our population. We would still need to recognise the importance of food security as breadbaskets become deserts, water tables fall and our own farm base dwindles. We would still need to address the fact that we are dependent on oil for our every need – a finite resource to which access can never be guaranteed. We would still need to prevent the destruction of forests, coral reefs, wilderness areas and the species which depend on them.

We would still, in other words, have a big problem on our hands. And we would still need to act swiftly and with determination to prevent it from getting worse.

It is often hard to reconcile the relentless horror stories with the reality of Britain today: a reality in which life, for many people, is materially better than it has ever been. Two centuries of industrialisation and economic growth have brought huge material progress. We have better homes, jobs, education and health care than ever before. We can fly to any nation in the world in a matter of hours. The Internet can find us almost anything at the click of a mouse.

But the global economy does a good job of hiding its consequences. It is a hugely effective system for delivering immediate wealth, but it grows at the expense of the natural world; its fresh water, forests, hydrocarbons, fisheries and farmland. The effect is that almost none of the wealth it creates can be transferred to our children.

Mass Migration

If even the most conservative estimates relating to climate change are accurate, we will see a wave of human migration on a scale we have never before had to accommodate.

In 2007, the Intergovernmental Panel on Climate Change predicted that:

> 'By the 2080s, many millions more people than today are projected to experience floods every year due to sea level rise. The numbers affected will be largest in the densely populated and low-lying megadeltas of Asia and Africa while small islands are especially vulnerable. Climate change over the next century is likely to adversely affect hundreds of millions of people through increased coastal flooding.'

The International Red Cross produces an annual Disaster Report. A recent one tells us that in the seventies, the number of people whose lives were affected by natural disasters was about 275,000. By the nineties that figure had jumped to 18 million – a 65-fold increase. The organisation also reports that 5,000 new environmental refugees are created each day.

And these are not just victims of the more shocking, visible disasters. Klaus Toepfer, ex-Director of the United Nations Environment

Programme (UNEP), has predicted that by 2010, the number of people on the move to escape the effects of 'creeping environmental destruction' will reach 50 million. The IPCC believes that number will swell to 150 million by 2050.

It's not hard to see how that might happen. Roughly 100 million people live in areas below sea level, and – given that the vast Greenland Icesheet is shrinking by eleven cubic miles each year – their prospects aren't good. The World Glacial Authority has told us that seventy-nine of the eighty-eight glaciers it has studied are retreating.

But it's not just people in low-lying lands who should be alarmed. 124 villages have been abandoned in recent years in Iran as a result of soil erosion. In China, the Gobi Desert is growing by 10,000 km2 each year. The UN calculates that worldwide – a quarter of a billion acres of good land are lost each year – affecting the food security of more than a billion people.

Environmental refugees now outnumber conflict refugees –as a result, there is a campaign afoot to require the international community to formally recognise their status. But we only have to look at the social tensions created by a few hundred thousand immigrants entering this country to understand that the movement of hundreds of millions of people across borders is simply not possible.

Finding a way forward

There comes a moment where the news is so bleak that people are inclined to throw their arms in the air and simply give up. Faced with a barrage of bad news in relation to the global environment, people increasingly ask, 'what's the point?' Even if Britain magically gets its act together, they say, what difference can that possibly make if other countries do not follow?

But while the problems are indeed vast, they are not insurmountable. Solutions exist, relatively straightforward, even painless ones. However they need to be on the same scale as the problems.

We cannot, for instance, simply 'green consume' our way to sustainability. We can buy energy efficient light bulbs and organic food; we can invest our money ethically, and growing numbers of people do. All of this is good news, but for this to make a real difference, they would need to be taken up by everyone, and realistically that just isn't going to happen in time.

It's not that people are uninterested in being part of the solution. Virtually every opinion poll on the subject shows that the majority of people genuinely value the natural environment. Time and again they express strong views on tackling climate change, protecting local landscapes and living sustainably. The trouble is, most green choices cost more. If you want to be environmentally friendly – drive a green car, take the train or eat good quality local food – the cost can be prohibitive. For many people, it's just not a realistic option. The challenge is to make it possible for everyone – not just the wealthy – to make green choices, and to save money doing it.

It is government leadership that will be the difference between success and failure. Unless pollution and the use of scarce resources become a direct financial liability, we have no realistic chance of shifting to a clean economy.

Politicians know this. The environment has never been so high on the political agenda. It has moved from being the preserve of professional environmental organisations into the public sphere. Global businesses like BP, Shell and HSBC write open letters to the Prime Minister calling for greater clarity on climate change policies.

But few politicians are prepared to take the necessary action. Nothing happens. Time ticks by, the situation grows more urgent – and government does nothing. Why?

Politicians are terrified of acting because they believe that tackling the looming crisis will involve restricting people's lives. They believe that saving the planet means inhibiting the economy, and that neither business nor voters will stand for it. They fear the headlines of a hostile media. They fear, ultimately, for their careers. It always seems easier to do nothing – to let the situation drift and hope that someone else takes the risk.

It's understandable, particularly in light of the recession. But the necessary changes do not need to be painful. Almost everything that needs doing is already being done somewhere in the world. If we took the best of today in every sector and made it the norm tomorrow, we would already be halfway there or further. And where companies, communities and even governments have done the right thing, they have been rewarded for it. Genuine solutions are there, and they are working.

Leadership, not headlines

There is an appetite for green solutions among the public, and it would be a tragedy if we allowed it to be extinguished as the result of clumsy green policy. Unfortunately, where politicians *have* taken steps, that is exactly what we've seen: contradictory policies, superficial attempts to grab headlines, or punitive measures that give green politics a bad name.

If a government is serious about the risks of climate change, it doesn't build homes on flood plains. If it is genuinely concerned about the growth in emissions from aviation, it shouldn't plan to treble airport capacity. If it knows that fifteen of the world's seventeen fisheries are at the point of collapse, it shouldn't make policy as if those stocks will last forever. If it wants to change consumer behaviour it shouldn't camouflage stealth taxes as 'green taxes' and enrage consumers and businesses alike.

The lack of real political commitment in Britain is increasingly leaving us behind. In Germany, Angela Merkel is driving her country towards rapid emissions reductions with targets far higher than EU or UK levels. President Obama has promised to invest £61 billion in a 'Green New Deal' to stimulate the US economy, create 5 million 'green collar jobs' and to accelerate the transition to a clean economy. Japan has said it intends to create a million such jobs, South Korea has promised to invest £38 billion in green technology by 2013, and Spain is planning to build 6,000 miles of High Speed Rail by 2020. If we are concerned about our 'global competitiveness', this is the competition we should be aiming to win.

That politicians now know the scale of the problems we face as a planet gives them a moral as well as a political imperative to take the necessary action. If they procrastinate, they do not deserve the privilege of government. They must either act, or let others act, but business as usual is not an option.

The tools for change

Our defining task is to marry the environment with the market. In other words, we need to reform those elements of our economy that encourage us to damage, rather than nurture, the natural environment.

The great strength of the market is its unique ability to meet the economic needs of citizens. Its weakness is that it is blind to the value of the environment. Unrestrained, we will fish until the seas are exhausted, drill until there is no more oil and pollute until the planet is destroyed.

But other than nature itself, the market is also the most powerful force for change. So we need to find ways to price the environment into our accounting system: to do business as if the Earth mattered, and to make it matter not just as a moral choice but as a business imperative. Destruction of the natural environment has to become a liability, not an externality. We shouldn't have to choose between the economy and the biosphere; we must merge them. That means rejecting growth based on environmental degradation, and rigorously applying the principle of making the polluter pay.

This is a fundamental principle. Put into practice, it would rapidly change the economy. Polluting companies would be at an economic disadvantage, and clean ones would be favoured by the market.

Today, the opposite is more likely. Dirty companies are able to offload the costs of their pollution onto the taxpayer, and regularly do. Global taxpayer subsidies to fossil fuels worldwide, for instance, are estimated to be in the region of £152 billion each year.

So what specifically needs to be done to reframe the way markets work?

Firstly, we need to use market-based instruments such as taxation. And when these tools cannot work, we need to change the boundaries within which the market functions by using well targeted regulation.

Taxation is the best mechanism for pricing pollution and the use of scarce resources. If the tax emphasis shifts from good things like employment to bad things like pollution, companies will necessarily begin designing waste and pollution out of the way they operate.

But people do not trust governments, so it's crucial that whatever money is raised on the back of taxing 'bad' activities, is seen to be used to subsidise desirable activities. For example, if a new tax is imposed on the dirtiest cars, it needs to be matched, pound for pound, on reductions in the price of the cleanest cars.

Green taxes have already earned a bad name in the UK, principally because wherever they have been introduced, they have been retrospective and set at levels that won't realistically change behaviour. And because the proceeds have not ostensibly been earmarked to subsidise green alternatives, they have quite rightly been seen as stealth taxes. A rare example of a green tax that both worked and was accepted by the public is the 1996 Landfill tax, which immediately transformed waste into a liability. The proceeds of the tax were initially reinvested into communities affected by landfill sites.

The other major tool in the policymakers' kit is trading. Carbon emissions trading is a good example of a market-based approach which attaches a value to carbon emissions and ensures that buyers and sellers are exposed to the price. As long as the price is high enough to influence decisions, it can work.

Finally, we also need a fresh approach to regulation. Direct controls force polluting industries to improve their performance, and can eliminate particularly hazardous products or practices altogether. Markets without regulation would not have delivered unleaded petrol, for instance, or catalytic converters. Without regulations requiring smokeless fuel, London's smogs would still be with us. Similarly, without new regulations, there can be zero doubt that we will exhaust the world's fish stocks.

Regulations are key. But the regulatory approach needs to be strategic. With some products and processes, the regulatory bar needs to be raised internationally to avoid companies chasing the lowest standards globally. And we need to shift away from an obsessive policing of processes towards a focus on outcomes. If the regulatory system is too prescriptive, there is no room for innovation, and no real prospect of higher environmental standards.

The alternative is an approach based on trust. The government should set high standards but not dictate how to meet them. By pulling back, assuming the best instead of expecting the worst, the government would be freeing farmers, traders, providers and businesses to innovate. This approach works, but only if the government has the strength to step in heavily where trust is abused.

Buying Power

The government has vast purchasing power; about £125 billion each year, and should use it to stimulate the market in sustainable goods and services, for example energy-efficient buildings, fuel-efficient cars, recycled products, brownfield instead of greenfield developments.

Of all the levers for change, this is the easiest and perhaps the most obvious. The Government knows this, and has made extravagant promises. But there has been little progress.

In 2007, the Director of the Government's own Sustainable Development Commission reported that:

'Overall, government performance is simply not good enough. Against a background of non-stop messages on climate change and corporate social responsibility, the government has failed to get its own house in order. It's absolutely inexcusable that government is lagging so far behind the private sector, when it should be leading the way.'

19

The ten steps

We know action is needed – and fast. We know what the problems are, and what mechanisms can be used to tackle them. Now we need the solutions.

Here are ten key steps any government must take if we are to navigate our way out of the current crisis and restore balance to our relationship with the world around us. This is a programme for action: a ten-point plan for turning Britain around and moving it in the right direction. It looks at the key environmental problems we face as a nation, and provides workable, practical solutions.

Put in place, these ten steps would help us do just that. They would help us achieve a constant economy; one where resources are valued, not wasted, where food is grown sustainably and goods are built to last. A system where energy security is based on the use of renewable sources, and where communities are valued as our greatest hedge against social, economic and environmental instability. They would deliver an economy that operates at the human scale, and above all one that recognises nature's limits.

Population

There is an issue that deserves a book to itself: population growth. If current predictions are correct, it represents a colossal challenge. It can even be described as our defining challenge.

Global population reached 6.8 billion in 2009, and is estimated by the United Nations to reach 9.15 billion by 2050. That increase – 2.3 billion – is roughly what our total population was in 1950. We are adding an extra 1.58 million people every week – the equivalent of a sizeable city.

Most of this increase will happen in developing countries, where any hope of genuine poverty alleviation is surely conditional upon the problem of over-population being addressed, and fast.

Pressure on the world's resources is already greater than can be sustained. If population trends continue, it is difficult to imagine humanity surviving the resulting environmental crisis. At best, we will see an increase in conflict as the competition for scarce resources reaches fever pitch.

But the difficulty is that there are no proven or ethically accept-able solutions to population growth. Restrictions on families will always be unpopular if not unacceptable. Appeals to people to forego children will likely fall on deaf ears.

The Optimum Population Trust reports that there are 80 million unplanned pregnancies a year – equal to the number by which world population increases annually. They believe this could be prevented by allowing full access to family planning worldwide. Such a policy may help, but it is far from certain.

For a country like Britain, which faces its own population pres-sures, there are domestic solutions. The Office for National Statistics projects that the UK population will pass 70 million by 2029. If that happens, it will result in immense pressure on our infrastructure and environment.

Two-thirds of the increase will be either directly or indirectly due to migration. Newcomers are currently arriving at the rate of about half a million a year – or nearly one a minute. Clearly we need a bal-anced approach so that immigration is brought into line with emigration.

The debate surrounding immigration has been fraught with ten-sion in the UK. Until recently, where people questioned the numbers, they were invariably accused of opposing immigration altogether, or worse, of being racist. The reality is that there are many people who can see the benefits of diversity and who believe a complete freeze on immigration would make this country a less interesting, less vibrant place to live, but who also believe that the sheer number of immigrants is greatly excessive. However, even if Britain takes action to stabilise its own population, it won't help the larger global issue.

The fact that there is no obvious solution should not inhibit discussion and debate. On the contrary, it needs to be raised far higher up the political agenda. At the very least, we need to recognise that population growth provides added reason for us to develop ways to live within our means.

We may find solutions, and we may even implement them. But more likely, if population is stabilised, it will be because of natural forces beyond our control, and not as a consequence of human policy.

Step One

Measuring What Matters

We have to abandon the arrogant belief that the world is
merely a puzzle to be solved, a machine with instructions
waiting to be discovered, a body of information to be fed
into a computer in the hope that sooner or later it will spit
out *a universal solution…*
Vaclav Havel

It's called the Information Age with good reason. We are inun-
dated with facts, figures, surveys and statistics. On 24-hour
rolling news services, on our mobile phones, our Internet
home pages and giant screens at major train stations, we are
constantly kept abreast of the most important stories affecting
our homes, our country and our world. Yet despite this
unprecedented dissemination of knowledge, we seem unable
to face up to some of the most dramatic and obvious challenges that stand in our way.

In 2006, the WWF published a damning report on the UK's
drain on the world's resources. If everyone on earth consumed
the same resources as the citizens of the United Kingdom, it
argued, we would need three planets to support us. Taking US
consumption it would be as much as five planets' worth. If
population doubles in forty years, as predicted, and people
everywhere consumed as we do today, we would need to

23

increase the level at which we exploit the natural world by a factor of sixteen. And if, as virtually every government hopes, we see an averge 3 per cent economic growth every year until the end of this century, global consumption of resources will increase by 1,600 per cent.

No rational person believes it is remotely possible to add this burden to the world, but the policies pursued by most governments assume it is. Why? To quote Bill Clinton: it's the economy, stupid. More specifically, it is the inability of our economists to measure what counts.

GDP: a 'grossly distorted picture'

Almost every nation on earth uses gross domestic product (GDP) to measure its economic growth. The trouble is, expressed as a monetary value, GDP simply measures economic transactions, indiscriminately. It cannot tell the difference between useful transactions and damaging ones.

So for example, if every man in Britain were to pay his neighbour for sex, we'd see a marked increase in the nation's GDP. Chopping down a rainforest and turning it into toilet paper increases GDP. If crime escalates, the resulting invest-ment in prisons and private security will all add to GDP and be measured as 'growth'. When the *Exxon Valdez* oil tanker ran aground and spilt its vast load of oil on the pristine Alaskan shoreline, US GDP actually increased as legal work, media coverage and clean-up costs were all added to the national accounts.

While GDP measures what we produce, it does not have the ability to factor in the cost of what we destroy to make it. It can only add – it can't subtract. We could empty the oceans of fish, chop down every last tree, fill the rivers with poison and pollute every last breath of air, and all the time, GDP could still be rising, and the economy could still be growing.

In other words, what most of us would regard as negatives, the economy measures as positives. Ultimately, GDP tells us nothing about the kind of country we actually inhabit.

Britain, for instance, has enjoyed an average yearly rise in GDP of 2.3 per cent since 1945. But has the quality of life for the majority of people risen by the same margin? In some areas, yes. Our homes are heated; we have access to decent medicine and health care; we have more gadgets than ever. But in other areas, the answer is an emphatic 'no'. Libraries are full of books describing a deepening 'social recession' in the UK. More than 2 million Britons, for example, are on antidepressants. A million regularly take class A drugs. Self-harming has reached record levels, as has binge drinking. The Samaritans say that 5 million people are 'extremely stressed'. British children aged 8-15, meanwhile, watch an average of 13.9 hours of television every week and are, according to UNICEF, the unhappiest in Europe. The fact that some 400,000 of them are on behavioural drugs like Ritalin, suggests that childhood itself has come to be seen as a disease.

Ironically, the man who helped develop the concept of GDP in the first place, Nobel Prize economist Simon Kuznets, never anticipated its use as a measure of progress. 'The welfare of a nation', he argued in 1934, 'can scarcely be inferred from a measurement of the national income.' Three decades later, US Senator Robert Kennedy said something similar: 'GDP does not allow for the health of our children, the quality of their education, or the joy of their play,' he said. 'It does not include the beauty of our poetry or the strength of our marriages, the intelligence of our public debate or the integrity of our public officials. It measures neither our wit nor our courage, neither our wisdom nor our learning, neither our compassion nor our devotion to our country; it measures everything, in short, except that which makes life worthwhile.'

But the pursuit of economic growth, as measured by GDP, has been the overriding policy for decades, with the effect that the consequences have often been perverse, and while economists have been telling us all the time that things are improving, reality has often been telling us the opposite.

For instance, when world leaders gathered in New Hampshire towards the end of World War Two to establish a way to rebuild the international economic system, they agreed that it would only be through world economic growth, world trade and the globalization of economic development that the poor would be lifted out of poverty.

That meeting saw the birth of the World Bank, the IMF and what eventually became the World Trade Organisation. The then US Secretary of the Treasury, Henry Morgenthau, declared 'The creation of a dynamic world economy in which the peoples of every nation will be able to realize their potentialities in peace and enjoy increasingly the fruits of material progress on an earth infinitely blessed with natural riches. . . That prosperity has no fixed limits. It is not a finite substance to be diminished by division.'

Since then, economic growth has expanded roughly five-fold. International trade has expanded by 6 per cent each year, and foreign direct investment has risen about three times the rate of trade expansion. In other words, the goals were met, and many times over.

But in the same period, we have seen the emergence of a situation where a fifth of the world's people now consume roughly four fifths of the world's resources, where more than a billion people live in urban squalor, where nearly 80 per cent of malnourished children live in countries whose food base has – according to plan – been redirected towards intensive agriculture for export. All this in addition to an unfolding ecological catastrophe that threatens the very survival of our species.

The means were undoubtedly achieved, but no one can pretend we are any closer to the given end. There is something profoundly wrong in the way we measure economic growth.

Measurements that mean something

With GDP as the principal mechanism for measuring it, what we call 'economic growth' is often nothing of the sort. Take rainforests, for example. Alive and healthy, the Amazon basin provides perhaps the most important service of all. It not only harbours millions of species, many as yet unknown to science, it acts in effect as the planet's air-conditioning system – regulating temperature and rainfall, absorbing carbon and cleaning the air. Without it, the world would be thrown into turmoil – and yet alive it has virtually no recognized value.

It is only as the trees are removed and transformed into disposable goods, and as the lands are usurped by soya plantations, that the forest begins to gain recognized value. Economists are trained only to see value in nature once it has been effectively cashed in, but is that really *growth*? Where in our accounts are the costs factored in? Where do we recognize the almost immeasurable value of the services the Amazon provides, without which much of South and Central American agriculture could not survive?

We need to find a new way to measure growth – real growth. As the American author Paul Hawken says in *Natural Capitalism*, 'we need to subtract decline from revenue to see if we are getting ahead or falling behind. Unfortunately, where economic growth is concerned, the government uses a calculator with no minus sign.'

A number of organizations have tried to assemble a new

27

tool for measuring progress. But the result is invariably a toolkit that is monstrous in its complexity and too impractical for any government to use. A neater approach would be for the government to establish a wholly independent Progress Commission, staffed by experts from a wide variety of fields: economists, environmentalists, statisticians, academics, etc. It would identify a set of key indicators for environmental, economic and social health – areas for which data can reliably be collected.

It would track signs of unhappiness like suicide and use of antidepressants, drug use and crime levels, as well as the amount of leisure time available to people to spend with their children. It would look at fish stocks, air pollution, biodiversity, energy security, food security. It would also look at living standards, income, health and access to public services. In all these areas, the commission would work closely with the Office of National Statistics, which would itself be insulated from government to ensure total independence. After all, no government in history has been able to resist the temptation to spin statistics to reflect well on their policies. Whichever indicators are selected, the results would be handed each year to Parliament and the media. The government would be required to respond.

The commission wouldn't comment on specific policies. By enabling us to judge the effect of government actions against these indicators, however, there would be an added pressure on governments to question their assumptions, and ultimately to craft policies on the basis of what really matters to people.

But beyond the measurements, how do we begin to put a realistic value on the natural world, and a cost on those things we don't want? For instance, how much is a forest, or a river worth? Clearly something. But what is the value of something we cannot do without? A stable climate, for example?

We can protect these assets using regulation, and for some problems that represents the only viable solution. Without regulation, we will surely deplete the world's oceans for instance. But by far the most effective, and powerful tool available to governments is taxation.

A tax effectively makes an activity or a product more expensive. On the flip side, a subsidy brings its price down. It is inconceivable that we will move to a sustainable economy without a radical programme of green taxation. That doesn't mean extra taxation. It means shifting the burden away from taxing good things like employment, and towards pollution, waste and the use of scarce resources.

If that happens, companies will necessarily begin designing waste and pollution out of the way they operate. Dirty companies would be at an economic disadvantage, and ethical ones would be favoured by the market.

In principle, the British government is already committed to 'green taxation'. But in practice the change has been negligible. The actual level of green taxation has fallen since 1997 from 9.4 per cent to 7.7 per cent, even while the tax take as a whole has soared. Roughly 50 per cent of the country's tax revenue still comes from income tax. The government's own Sustainable Development Commission has described its use of the tax system as a 'significant failing'.

The overwhelming bias in the current tax system is for indiscriminate economic growth, with among other things vast tax breaks on fossil fuels.

We need a major shift in the tax bias, one that is dramatic enough to change behaviour and the way we do business. Whatever taxes are set must be set realistically, and within a realistic timescale. There is no advantage in crippling businesses. Equally, our approach must be bullish and brave. But what is key to all green taxation is a recognition that people simply do not trust governments. So wherever a tax is levied against a 'bad' activity, it must be used to bring down the cost of 'good' activities. Any other approach will rightly breed mistrust and resentment.

There will always be people who reject the very notion of using tax as an instrument of change. But tax cannot be anything other than an instrument of change. Indeed it's hard to think of any existing tax that does not have some kind of impact. Taxation is neither neutral nor objective.

And if anyone is in any doubt about the potential in green taxation to bring about change that works both for the environment and business, this book is full of specific examples of ideas that not only will work, they already do somewhere. Examples of countries that have taken the initiative; and of companies that have sought to clean up their performance and have seen huge savings as a result.

We don't have to choose between the economy and the environment. In a constant economy, the two are linked. Our challenge, then, is to reconcile them.

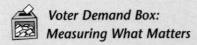

Voter Demand Box:
Measuring What Matters

1. Establish a fully independent Commission, staffed by experts from a variety of fields: economists, environmentalists, statisticians, etc. It would identify a set of key indicators for environmental, economic and social health – areas for which data can reliably be collected. Working with a wholly independent Office of National Statistics, the commission would conduct an annual audit, and require the government to respond. People would be able to judge the effectiveness of government policy on areas that matter to them.

2. We cannot move to a sustainable economy unless we learn to put a value on natural capital. We need to include our natural wealth in the national balance sheet, and to measure its loss or depletion as a cost.

3. A radical programme of green taxation. That doesn't mean extra taxation. It means shifting the burden away from taxing good things like employment, and towards pollution, waste and the use of scarce resources. Green taxes should never be allowed to become stealth taxes.

Step Two

Power to the People

I am not one of those who think that the people are never
in the wrong. They have been so, frequently and outra-
geously, both in other countries and in this. But I do say,
that in all disputes between them and their rulers, the
presumption is at least upon a par in favour of the people.

Edmund Burke, 1770

British parliamentary politics is in a depressing state. More
and more people feel they have less and less control over the
political process; consequently many have less and less interest
in being involved in it. Rarely have politicians been trusted by
the public, but never have their reputations been so low.
Turnout in general elections has declined significantly and
allegiance to political parties has dropped to 14 per cent. The
last two elections had the second and third worst turnouts
since 1900. Only 1918 was worse – and that was amid the
chaos of demobilization after World War One.

The anecdotal evidence for this growing disillusionment is
equally depressing: radio phone-ins fizzle with rage and
contempt for the political classes; election canvassers report
weary cynicism on the doorsteps; 'politician' stands alongside
'traffic warden' and 'estate agent' as a profession it is almost
mandatory to despise.

You'd be forgiven for thinking that we didn't care, that the sum total of political thought in Britain was reserved for 'they're all as bad as each other' cynicism, or extremist bombast from dubious parties like the BNP which thrive on the public's disaffection.

But when the Commission on Parliament in the Public Eye warns that the UK is now very close to the point where a government could not claim democratic legitimacy, it's time to look long and hard at the society we have become.

Politicians can come up with gimmicky ideas to 'engage' young people like voting online or by text message, as if democracy were a series of *The X Factor*, but there remains an underlying assumption that we don't care, that politics is no longer important to us. This dismissive attitude misses the point entirely – we are, in fact, a highly politicized nation.

The fact that people don't engage in party politics in the numbers that they used to, does not make Britain an apathetic nation. A million people marched in London against the war in Iraq. Half a million people took to the streets in opposition to the ban on hunting. Millions of people are involved in community and charity work, and in 'single issue' pressure groups. It's not that we don't 'do' politics, we just do it in a different way.

The reason that fewer and fewer of us turn out to vote is because politicians have turned democracy in Britain into a superficial and disengaging process.

At the national level, politics is increasingly seen as a power game restricted to remote elites. For the 1,500 or so days in between general elections – when people can choose between two political parties with whose views they almost certainly do not agree in their entirety, and which are in any case increasingly similar in outlook – we are denied any access at all to the decision-making process. Once the polls have closed, we have no choice but to accept one often bad decision after another.

If anything the situation is worse at local level. Over the last two decades, local government has been denuded of its powers by governments of both main parties. Day after day, elected local representatives are overruled by central government on local issues, to such an extent that people wonder why they should bother voting in local elections at all. We have an absurd situation today where councillors elected on a specific platform – opposing for example a new incinerator – are then unable to vote on that issue because they are deemed to be 'prejudiced', despite the fact that they were elected to fight that very issue. The centralized nature of our politics means that local elections are not seen as important in their own right, but more often as a barometer of the national standing of the parties. For some, even that doesn't matter – as true power resides elsewhere: in Brussels. There, decisions are often made by people who are insulated from any democratic pressure at all. If poor decisions are taken there is little, if anything, the electorate can do about it.

The overall result is a feeling of helplessness, a sense that power is not in the hands of the people, but in the hands of unconcerned and unrepresentative cabals. If the government decrees that genetically modified foods are safe and conse-quently removes what little protection exists for the British consumer – well, we have no choice but to accept it, even if science and public opinion say we shouldn't. If a supermarket is granted permission to build a large out-of-town hyper-market, there is absolutely nothing that local citizens can do about it. From motorway construction to 'faith schools', the removal of civil liberties to the banning of popular pastimes – if the people don't like it, there is very little they can do about it.

Democracy Makes You Happy

Countries that offer strong political freedoms and civil liberties tend to report higher levels of well-being. There is a growing body of evidence pointing to the crucial importance of citizens having adequate means to express democratic freedoms and, in particular, the opportunity to influence the decision-making that affects their everyday lives.

The practice of direct democracy is common in Switzerland, which frequently uses referenda in major decisions taken by the twenty-six different cantons that comprise the country's federal structure. Cantons vary in the thresholds set to trigger a referendum – for instance the minimum number of signatures required. Research has shown that those cantons with the greatest emphasis on direct democracy also register the highest levels of well-being.

Direct democracy has benefits for people who participate in referenda as well as for the Swiss public generally. A survey comparing budget decisions in over 130 Swiss towns found that public expenditure and debt are lower in cities that require a budget deficit to be approved by citizens. Cantons with direct democracy also report higher overall productivity and lower corruption, and their citizens are better informed about the issues concerned.

A new approach

Britain's parliamentary democracy was already well established 200 years ago and remains more or less intact today. The franchise was gradually extended until it became universal with the granting of women's suffrage in the earlier part of the last century, but since then, reform has been comparatively minor, with the exception of devolution.

Parliament might not have changed, but society has. In the nineteenth and early twentieth centuries most adults were uneducated, uninformed and had little experience of the world beyond their locality. That world no longer exists. The talent gap between the ruler and the ruled has shrunk. The public today has unprecedented access to information. Most school leavers go into further education. Almost every aspect of our lives has been transformed by social, economic and technological change, but the way we make collective decisions remains stuck in the past, hopelessly outdated. We are still expected to hand over political choices to an exclusive group of professional politicians who cannot meaningfully be held to account, except on one day, twice a decade.

We need to rethink the relationship between people and power, and to develop a model of political citizenship that is appropriate for our times. We need to undertake radical and urgent reform of our political system – none of the limp suggestions we have so far seen from mainstream parties are anywhere near enough. We need the equivalent of a new Great Reform Act, to galvanize the people and rejuvenate democracy.

The principle on which this new wave of reform should be based is a simple one: localism. Power needs to be diffused throughout the community. The wider and more authentic the levels of decision-making, the more people who become involved and the healthier the society. There are some things that, by their nature, can only effectively be done together with our international partners. Nothing, however, should be done internationally that could be done better in Westminster. And nothing should be done nationally that can be done locally.

'Lift the curtain' Enoch Powell wrote, 'and the state reveals itself as a little group of fallible men in Whitehall, making guesses about the future, influenced by political prejudices and partisan prejudices, and working on projections drawn from the past by a staff of economists.'

It is time to place less faith in the power of big government, and to recognize that a sustained change in values can only be built from the bottom up, underpinned by a broader consensus in society than exists today.

The Barnes Vote: People Power in Action

Shortly after being selected as a parliamentary candidate for Richmond Park and North Kingston, I was contacted by a local campaign group that was trying to prevent the opening of a new Sainsbury's supermarket.

In 2007, the company had applied to build a new store in a much-loved shopping area of Barnes. The council was bombarded with letters of protest from residents who were concerned about increased traffic and the effect on the local independent shops they had grown to love. The council rejected the company's application, but was swiftly overruled by the National Planning Inspectorate in Bristol – over a hundred miles away from Barnes.

In protest, the White Hart Action Group was formed by local resident David Rossiter, and at a packed public meeting it was decided that the residents would commission a professional, independent referendum on the future of the site. On their behalf, I contacted Electoral Reform Services, and we set a date. Sainsbury's was invited to review the question and to include in the referendum package a letter from them to residents.

A month later, the referendum was held and the turnout was greater than in the previous general election. Eighty-five per cent of people voted against the new store, and as an expression of people power, the point was made. However, the referendum had no teeth in law, and has, so far at least, been ignored by Sainsbury's.

The referendum directly answers the charge that people don't care. It also shows the exaggerated power of large corporations, and the full extent to which local people have lost control of local politics.

Total Recall: The Birth of the Governator

Under Californian law, voters are able to initiate a process where politicians can be 'recalled' from office. If enough signatures are collected, voters are presented with a ballot asking them first if they believe the politician should be 'recalled', and in the event that a majority answer 'yes' they are asked who they want him/her to be replaced by.

The process has been tried on a number of occasions. California governors Pat Brown, Ronald Reagan, Jerry Brown and Pete Wilson all faced unsuccessful recall attempts. However, in 2003, voters successfully recalled sitting Governor Gray Davis.

The campaign spanned the summer of 2003. After several legal attempts failed to stop it, California's first-ever gubernatorial recall election was held on 7 October, and Davis became the first governor recalled in the history of California. He was replaced by action movie hero Arnold Schwarzenegger.

California however provides stark lessons in how direct democracy can be abused. For instance there are no proper limits on pre-election expenditure, which means that referendums in California, like elections throughout the US, cost millions of dollars and are beyond the reach of the people the system was designed to serve. As a result, referendums have become just another avenue for vested interest groups. Indeed a specialized industry has grown up around the process, with companies providing expensive services such as signature gathering and campaigning. The situation is compounded in California by laws requiring two thirds of both houses of legislature to approve budgets and tax increases. The effect is that minority parties have veto power at budget time, and there is often a stalemate. Direct Democracy is hugely popular in California, but there is a general recognition that the system needs reform.

It's time for direct democracy. What that means, very simply, is that ordinary people are given significantly more power – real power – to intervene on any political issue, at any time of their choosing. With sufficient popular support, existing laws can be challenged, new laws can be proposed, and the direction of political activity, at local and national level, can be determined by people rather than elites.

This would radically transform politics. Not only would we be able to stop many unpopular policies from becoming law; we'd also be able to kick-start positive changes. The whole process of calling a referendum would ensure more widespread and much better informed debate.

We'd also see greater legitimacy given to controversial decisions. Under the current system, it's difficult not to feel swindled or hoodwinked when a policy on which you believe you hold the majority view is defeated by votes in Parliament or in the council chamber. Under direct democracy the losers at least have the important consolation of knowing that they were given the opportunity to make their case to their fellow citizens on a level playing field.

It's not a new idea. Direct democracy is already used in other, more responsive democracies around the world, including notably at state and local level in the USA. These rights would apply at local and national levels and would help strengthen dialogue between the governed and governors, empowering those sections of the community who feel their voices are not heard. Concerns that those with financial backing could skew the agenda can be addressed by placing a limit on expenditure, just as applies for general-election campaigns.

It is vital that any new mechanism for giving local people greater ownership of decisions is genuine, and seen to be so. The tendency of politicians to concede the principle of power-sharing while attempting to retain control of decisions in practice has done much to deepen cynicism and disengagement from the political process. Politicians must face up to the reality that the popular view of them is one mired in mistrust. Any reform must be real rather than cosmetic, substantive rather than a gimmick.

Revolution in Brazil: The Participatory Budget

In 1989, the Brazilian Workers Party (PT) – whose then leader, Luiz Inácio Lula da Silva, is now the country's president – instituted a radical shake-up of the spending process in the city of Porto Alegre, which they controlled. They removed the power to decide on the city's budget from the usual councillors and technocrats in city hall and began a process of popular consultation, which has evolved annually to iron out glitches that appeared along the way. The idea was – and remains – to allow the people of the city to decide how their money is spent.

The city's budgetary priorities were divided into themes – environment, transport, culture, health, education and so on – and for each one they held a regular public conference in each region of the city. Any citizen could participate in these conferences, which debated how much money should be spent on the theme. Each conference then elected a delegate, who put forward the decisions made by their citizen electors. The proposed budget could only be put into practice if approved by the people.

The Porto Alegre Citizens Budget was a huge success. People felt they had a role in deciding where their money went – and because they were forced to make hard choices about spending priorities, they gained an appreciation of the importance and the reality of local politics. Since it was introduced, Porto Alegre's people have used their money to pave 25,000 km of roads, provide 96 per cent of homes with clean running water, steadily increase sewage provision, set up family health clinics, work towards eliminating child work and expand the number of schools. The money is spent on what people want – not what politicians think they want.

The case against direct democracy doesn't add up

Direct democracy – and specifically the use of referenda – has fierce critics, though their objections rarely stand up to scrutiny. The oft trotted-out line that the public is too irresponsible to be trusted with important decisions is an argument against democracy itself. Indeed, it has been used since the early days of Parliament to oppose every extension of the franchise, including votes for women.

Critics also point to the newspapers and other media, warning that they would wield an unhealthy influence over the country. In reality, newspapers have far more influence over 650 MPs than they ever could over a notional audience of 60 million. In recent years every government has quailed before powerful newspaper editors, and most MPs are only too keen to court influential journalists. The penalty for falling foul of a newspaper can be severe. With an unblackmailable electorate of millions calling the shots, direct democracy would actually reduce the power of newspaper proprietors and editors to impose their agendas on fearful governments and MPs – and the same can be said for the shadowy special- interest groups that some believe would use their influence and financial muscle to ensure a referendum went their way.

If there is one criticism that might give pause, it is that legislation is often complex. But when we look at the way government already deals with such decisions, we can see a ready-made solution. The House of Commons already subcon-

41

tracts detailed scrutiny of a bill that comes before it to a committee stage where a smaller group of MPs examines it in depth. Eventually there is a final vote on the bill in each House of Parliament. Obviously, the bulk of legislation would never be subject to direct democracy but if the matter was of sufficient public interest there is no reason why a referendum could not be held after these final votes, but before the royal assent.

What's more, a referendum, even one dealing with a complicated subject, would prompt precisely the kind of public engagement that politicians are desperate to encourage. Knowing that their vote would have an impact on the future would bring out the best in people and raise the quality of debate, often with surprising results.

The greatest fear that irresponsible opinions would always win the day is not borne out by practical experience. Many states in the US have direct democracy – so-called propositions – and the results betray no overall ideological direction. For example, California voters have backed the medical use of marijuana, and opposed relaxation of restrictions on gambling.

Perhaps the most telling recent example of the ideological unpredictability of referendum results happened outside of the United States. In June 2008, a nation with a reputation for insularity, was asked to tighten its citizenship laws, making it harder for foreigners to gain naturalization. Much to the surprise of international commentators, the proposal was rejected by a margin of almost two to one. This country can justly claim to be the most democratic on earth: Switzerland.

Making it work in Britain

In Britain today we have a well informed and educated population. It is time to place more confidence in it – in us – to act responsibly by giving people a greater sense of control over decisions that influence the quality of everyday community life.

Direct democracy is a principle that resonates throughout a nation's political life. The key principle, however, is that decisions are always taken at the lowest possible level. As an example: if there is a proposal to build an incinerator in a particular borough, people living in that borough would be able to 'earn' the right to hold a referendum if they manage to collect a specified number of signatures. If the issue is still more local – a proposed supermarket, say – then the referendum would be held at the level of the ward. The same process would apply, only at a more local level still.

We would, of course, need debate about the kind of issues that could be influenced, made or reversed via referenda nationally. Issues relating to war, for instance, would need to be exempt, as an elected government would need to be able to make speedy decisions. Constitutional issues, like the transfer of powers to the EU, though, would justify use of a national ballot initiative.

Taken together, these proposals could transform democracy in Britain, bringing our constitution into line with more enlightened countries, and giving people a real say in how their localities are run, and who runs them. Trusting people to use the political power which is rightfully theirs would help transform attitudes to politics, bring home a sense of responsibility and opportunity and help to make the UK a more hopeful, and functional, society.

If we improve the process of decision making, we will improve the quality of decisions made. Politicians alone will always be reluctant to take risks. But through our collective wisdom, we will take decisions that suit our collective interests. Parents will vote with their children in mind. Residents will vote with their neighbourhoods in mind. Pure democracy is a prerequisite to restoring balance.

Switzerland: Direct Democracy in Practice

In Switzerland, if 100,000 signatures can be collected within an eighteen-month period, then a proposal can be proactively put on the ballot paper and voted on by the general public. If it is passed, then it becomes law.

One example of this in practice is the so-called 'Alpine Initiative'. In 1987, the region of Uri was badly damaged by violent storms. As a result, the transit roads had to be closed. Noticing the beneficial effect this had on the quality of the local air, the people of Uri decided that they wanted the roads to remain closed. So they began organizing a referendum to demand that all heavy goods should be carried by rail (with the exception of goods loaded or unloaded in Switzerland), and that there be no further expansion of the alpine transit roads in their region.

Despite heavy pressure from the Federal Parliament, the National Council and the Council of States, the united people of Uri won their battle, and the motion was approved in 1994.

Transition Towns

The Transition Towns movement is a powerful example of ordinary people taking control at a local level. Internationally there are 500 such initiatives.

One of the most exciting is Kinsale in Ireland, where the town council has accepted plans to work towards energy independence. Other initiatives include community gardens, a proposal to establish the first community composting scheme, energy audits for domestic, commercial and municipal buildings, pedestrianization of part of the town, a prototype anaerobic digester for the town's commercial food and agricultural waste, and a 'free cycle system' in which people pass on unwanted belongings for free to others instead of dumping them.

Transition Towns have arisen in response to the big issues of the day: climate change, peak oil, food security. The communities are embarking on a programme that involves bolstering local food links, developing clean energy, pursuing energy efficiency and attempting to minimize the amount of waste they generate.

 Voter Demand Box:
Direct Democracy

With sufficient popular support, people should be able to challenge existing laws and propose new ones.

The mechanisms of direct democracy are many, and the precise formula would need to be debated. But three key mechanisms are:

- Ballot initiatives, where new laws are proposed by citizens.
- Popular referenda, in which existing laws can be challenged.
- Recall initiatives, allowing people to remove unpopular public officials.
- It should also be possible to trigger referendums and recall initiatives at a local level, for example the Borough or even the Ward. The key is that decisions are taken at the appropriate level.

'Our Say', one of the growing number of campaigns backing the greater use of referenda, has set out how direct democracy might work in practice:

- Each year, on Referendum Day, people would be able to vote on issues of concern, both national and local.
- To trigger a referendum on a particular topic, an agreed percentage of the electorate would need to sign a petition. If it were 2.5 per cent, a million signatures would be required to trigger a ballot on a national issue. For local issues affecting, say, a district council, this would require around 4,000 people to back the proposition.

- The Electoral Commission would check the validity of the petition and agree the wording of the question on the ballot paper to ensure that the question was fair and balanced.
- People would need to sign petitions in person and the signatures to trigger a vote would need to be collected in a one-year period.
- There would be strict limits on the amount of money that could be spent on referendum campaigns.
- Balance in TV and radio coverage of the issues under discussion would be a legal requirement, as well as fair access to other media coverage for each side.

Step Three

The Precautionary Principle

Never, no, never, did Nature say one thing,
and Wisdom say another.
Edmund Burke, 1797

Scientific endeavour has led to some of the most remarkable discoveries in human history. Advances in all disciplines have made us healthier, wealthier and able to live longer, fuller lives. So much so, in fact, that in the twentieth century, people began to look to science as reverentially as they once looked to God – 'Science is a new religion, and disinfectant is its holy water', as George Bernard Shaw once remarked.

Just as the devout worship their God with unwavering certainty, we have been encouraged to embrace science with the same kind of piety. Whether it's GM food, controversial vaccines, radiation, or fluoride in our drinking water, society is expected to accept, unquestioningly, scientific judgements. To question such edicts is to court derision, to stand up to the self-appointed experts is to run the risk of being branded a Luddite.

Over the last few decades, however, the fallibility of scientists has become more visible. High-profile cases like the thalidomide scandal of the sixties – when the Merrill Company claimed there was 'no positive proof' of a connection between the use of their drug during pregnancy and malformations in the newborn – have seen science's public profile take a significant downturn. It's not that we reject it, just that we realize it isn't always right all of the time.

As scientific debate has become more public, we have begun to understand the business of science more clearly than ever before – especially the way it is funded. The vast majority of scientific research is paid for by big business, for commercial ends. There is, of course, nothing wrong with this – in fact it's a necessity for advances to be made – but that doesn't mean regulatory bodies or the general public shouldn't look cautiously at the results.

It's a question of interests. A company engaged in developing GM plants has little incentive to look for potential hazards. Any problems found in development will slow down the process of bringing a product to market. In a highly competitive world, businesses regard science as something of an arms race, and speed is of the essence.

As big businesses hold the purse strings, there are far fewer grants available to examine the safety of new products than there are to approve them. The interests of consumers are therefore very much secondary to the swift return on investment. That, combined with the fact that a revolving door exists between the regulators and the regulated means that consumers are often right to be sceptical of official assurances of 'safety'. When runaway corporate science combines with a weak and sometimes corrupted regulatory system, consumers are invariably at risk.

After the biotech company Monsanto was caught boasting of its success at influencing the composition of UN food safety committees, the Canadian Royal Society warned: 'The public interest in a regulatory system. . . is significantly compromised when that openness is negotiated away in exchange for supportive relationships with the industries being regulated.'

This conflict of interests is perhaps best illustrated by the debate surrounding genetically modified food.

Genetic roulette

When asked to defend the safety of their products, the GM industry often points to the US, where millions of people have been eating GM with no ill effects. But how do they know? There have been virtually no studies made into the health effects of GM food.

GM food might be perfectly safe, but *might* isn't good enough. The industry claims that there are no signs of frightening new diseases. True, but there are countless examples of relatively common, but nevertheless unpleasant, illnesses on the rise. It is known, for instance, that more than 1 per cent of 3-year-old American girls show signs of puberty. Could that be linked to GM soya-based infant formula which has raised levels of oestrogen? We don't know, and without rigorous studies we will never know. Incidences of food-related illness in the US have doubled since GM was first introduced. Is there a causal link? Again, we don't know because the regulators haven't bothered to find out.

GM enthusiasts describe genetic engineering as a mere extension of the kind of plant breeding that has always driven agriculture. But in traditional plant breeding, similar strains are interbred and either thrive or die. If the strains are too distantly related, the experiment is not possible. With GM technology, genes are able to leap previously impossible natural barriers. There is no natural way to splice fish genes with cabbages to boost frost resistance. Genetic modification is the only way to achieve this.

But it is not a precise technology. When a gene is removed

49

from its context, its effects are unpredictable. That is why inserting alien genes into organisms leads to clear abnormalities in roughly 99 per cent of cases. According to the University of California there have been no studies into the less obvious defects, so it's impossible to know what percentage of the remaining 1 per cent are actually healthy.

Our lack of knowledge is due to a quirk of legislation: the GM industry is not required to invest in the kinds of studies expected of other industries. Pat Thomas, former editor of the *Ecologist* explains:

> Although hardly grounded in sound science, the fact that we may all already be eating GM food is the core of the 'safety' argument put forward by big biotech and the government. In any other field we would base our understanding of the safety of a thing in part on human studies. There would be toxicological assessments, tissue studies and long-term multi generational studies to identify any damage that accumulates in children, grandchildren and great-grandchildren.
>
> Unlike other large industries, such as those making mobile phones and pharmaceuticals, the GM industry is not required to invest in these studies. Instead, it is allowed to rely on a method called 'substantial equivalence' – described in the journal *Nature* in 1999 as a 'pseudo-scientific concept' that was 'created primarily to provide an excuse for not requiring biochemical or toxicological tests.' Using this method, just a few key chemicals – such as nutrients and known toxins – are compared to those in the non-GM plant. If the levels are considered similar, the whole plant is considered to be 'substantially equivalent' to its non-GM counterpart.

Despite this, there have been numerous reports from independent scientists on the dangers associated with GM food. In a 2005 Russian trial in which female rats were fed Roundup Ready soya before mating, during pregnancy and lactation, 56 per cent of the pups born to mothers on the GM diet died within three weeks of birth, compared with just 9 per cent of those whose mothers ate a non-GM diet. The surviving GM-fed pups had stunted growth and smaller organs.

Disaster Scenario

A few years ago the US Environmental Protection Agency granted a German company approval to begin testing a genetically modified soil bacterium at Oregon State University.

Designed to break down waste vegetation and produce ethanol as a by-product, it was a tremendous success. But when students added the processed waste to normal, living soil, and planted seeds, there were unexpected results. The seeds sprouted, but immediately died – all of them.

The GM bacterium had out-competed soil fungi, essential to plant growth, and rendered the soil effectively dead. Worse, the students discovered to their horror that the bacterium could survive and replicate.

According to David Suzuki, Canada's pre-eminent geneticist, 'the genetically engineered klebsiella could, theoretically, have ended all plant life on this continent. The implications of this single case are nothing short of terrifying.'

All this happened under the watchful eyes of the Oregon students. But had they not done their research properly, the bacterium would have been approved for commercial use, with unthinkable consequences.

Two UK trials funded by the Food Standards Agency – one, the only known trial involving humans eating GM soya; the other involving sheep eating GM maize – found that some of the genetic material remained in the gut after ingestion, and was transferred to gut bacteria. Monsanto's own research has found that rats eating GM maize developed smaller kidneys and showed startling changes in blood chemistry, including an increased white blood-cell count – indicating an immune response to the food.

At the very least, these studies should stir up a heated and open debate. But it's an argument that the British government is simply not prepared to join. Despite public misgivings and scientific doubts, the government remains resolute – GM technology is perfectly safe. So what is it that makes them so certain?

Their confidence is born out of a determination to push ahead with GM technology, ignoring anyone or anything that questions its chosen policy. They have invested heavily – both financially and politically – in the GM industry, and a U-turn at this stage is unthinkable. But there's another explanation for this bullish behaviour: infiltration of the regulatory system.

A glance at the career of the late Sir Richard Doll shows the corrupting impact just one person can have on the regulatory system.

A Pillar of the Cancer Establishment

About ten years ago, the *Ecologist* magazine ran a lengthy feature on one of the UK's most respected cancer experts, Sir Richard Doll.

In reviewing his track record, the magazine showed that 'In every field in which Doll has been involved he has systematically defended the interests of industry and state, even when these are in total conflict with those of people in general, and are irreconcilable with all the established knowledge on the subject.' The only possible explanation, the article concluded, was that he was being paid by industry.

His friends were appalled, and he himself contacted his lawyers. But after a short while he told the *Daily Telegraph* that he wouldn't be taking any action against 'a child's fiction magazine'. He added that he wouldn't respond to the *Ecologist*'s allegations because they were 'beneath contempt'. However, on his death in 2005, Sir Richard Doll was spectacularly exposed for being on the payroll of companies whose products he was supposed to be reviewing in the public interest.

When Doll submitted an unsolicited letter to a Royal Commission set up to establish the health effects of Agent Orange on Australian soldiers who'd fought in the Vietnam War, the effect was described as electrifying. He must have known that his defence of Agent Orange would have been less electrifying had he mentioned the fact that he was being handsomely paid by Monsanto – the makers of Agent Orange – at the time.

In 1957 Doll had cautiously suggested a quantitative relationship between radiation and leukaemia. But then in 1987 he looked at cancer rates in the vicinity of all Britain's nuclear power stations and concluded that there was 'no increase in childhood leukaemia near any nuclear power station' – even though the government had already accepted the existence of cancer clusters around nuclear plants in the same year.

Two years later Doll was again asked to assess the cancer risks associated with living close to a nuclear power plant. This time the results were less convenient. The death rates were found to be 21 per cent higher than the national average. Instead of drawing the obvious conclusion, Doll admitted that 'Until we find some other cause, we cannot say that radioactivity is not responsible.' Shortly after, he found his alternative: a 'leukaemia virus', for which to this day there is no evidence. British Nuclear Fuels backed much of Doll's work on radiation and cancer.

When Doll was paid by Dow Chemicals and ICI, he oversaw a review that largely cleared vinyl chloride – used in plastics – of any link with cancers. The WHO strongly disagreed with Doll's findings, but his evidence was nevertheless used to defend continued use of the chemical for more than a decade.

Perhaps Doll's worst legacy was his impact on the policies and direction of Imperial Cancer – or Cancer Research as it has become. Because Doll steadfastly refused to accept that environmental contaminants played a role in the increases in cancer rates, hardly any of the group's budget was ever allocated to identifying environmental causes of cancer and campaigning for their elimination.

'One of the biggest myths in recent years is that there is a cancer epidemic caused by exposure to radiation, pollution, pesticides and food additives,' a report by the organization said. 'The truth is that these factors have very little to do with the majority of cancers in this country. In fact, food additives may have a protective effect – particularly against stomach cancer.'

Doll himself warned against 'A large and powerful lobby against pesticides, which they say leave cancer-causing residues in our food.'

Doll made a name for himself in the fifties when his research firmly established the link between smoking and cancer. But because the link between smoking and lung cancer is so well established, it has become convenient for other industries to blame virtually every form of cancer on smoking. Doll wrote that cancer increases of late 'Can be accounted for in all industrialized countries by the spread of cigarette smoking.' But of the dramatic rise in cancer incidences since 1950, 75 per cent have been in sites other than the lung. And among non-smokers, the incidence of lung cancer has more than doubled in the past few decades.

Revolving doors

According to the European Commission there are more than 15,000 lobbyists currently operating in Brussels, almost all of whom are paid to fight for the interests of big business. They operate without restriction, and they enjoy extraordinary access to decision-makers, particularly within the commission. Their impact on policy is vast, as became dramatically evident in 1998 when the EU began developing a chemicals policy known as REACH to increase the safety of human health and the environment.

Its purpose was to identify and phase out the most hazardous chemicals by requiring their substitution with less harmful alternatives. Under REACH proposals, new chemicals would only be allowed on the market after rigorous testing. The burden of proof was to be shifted away from governments and onto the manufacturers themselves, who would also have to provide detailed information on all their chemicals to anyone coming into contact with them.

That was the idea. But what followed was described as the biggest lobbying exercise in history, and what emerged was a diluted, toothless and ineffective policy.

When the first drafts were circulated, the powerful European Chemical Industry Council began by denying there was even a need for REACH. In 2001, it claimed in a letter to MEPs that 'There is little direct evidence of widespread ill health or ecosystem damage being caused by the use of man-made chemicals.' It later dropped this argument, but continued to lobby all the key decision-makers in Europe.

A defining moment was when Tony Blair, Gerhard Schröder and Jacques Chirac issued a joint letter to the EU Commission, claiming that REACH would undermine the international competitiveness of the European chemical industry. The US chemical industry applied even more pressure. The American Chemistry Council boasted that it successfully 'Rallied opposition to the draft proposal, including a major intervention by the US government. . . These efforts. . . brought about significant concessions in the draft.'

So aggressively did the US government intervene on behalf of the US chemical companies that in 2003, more than seventy non-governmental organizations, public health professionals, nurses, environmental and community groups wrote a letter to President Bush asking him not to use government funds to undermine REACH. This was followed in April 2004 by a congressional investigation into the lobbying efforts which criticized the administration for relying almost exclusively on information supplied by the US chemical industry.

The German chemical industry meanwhile, in its 2005 annual report, boasted that 'Many members of the European Parliament have taken up our proposals. The rapporteur in the EU Parliament Committee for the Internal Market and Consumer Protection has largely accepted our proposals and presented them in the debate as a practicable alternative to the commission's proposed regulation.'

But it's not just corruption of the regulatory system that has allowed the rise of reckless and unsound science.

Arrogant science and nanotechnology

The Liberal Democrat peer and chairman of Sense About Science, Lord Taverne, believes science should be freed from the constraints of democracy, and describes the precautionary principle, where new products are regarded as guilty until proven innocent, as the 'cowardice of a pampered society'. He's not alone.

Sense About Science?

Every few months, newspapers run stories about celebrities 'getting their science wrong'. The source of the stories is usually an organization called Sense About Science.

The group produces a pamphlet that is full of what it regards to be false assertions by celebrities about the benefits of homeopathy and so on, and ends with an offer by the organization to act as a fact-checking service.

However, the group's own scientific pronouncements, and indeed the organization itself, have come under serious question.

For instance, it was recently denied funding by the respected Wellcome Trust which warned that the group 'runs the risk of being seen to be fuelled by "assumptions", and not direct evidence.'

Sense About Science is much more than an innocent fact-checking service. It is part of a bizarre political network that began life as the ultra-left Revolutionary Communist Party and switched over to extreme corporate libertarianism when it launched *Living Marxism* magazine in the late eighties. *LM* advocated, among other things, lifting restrictions on child pornography.

During the nineties, *Living Marxism* campaigned aggressively in favour of GM food. In 2000, it was sued for claiming that ITN had falsified evidence of Serb atrocities against Bosnian Muslims, and was forced to close. It soon reinvented itself as the Institute of Ideas, and the online magazine *Spiked*.

The chairman of this movement's latest incarnation, Sense About Science, is the Liberal Democrat peer, Lord Taverne. While he routinely fires off about non-scientists debating scientific issues, calling at one point for Prince Charles to be forced to relinquish the throne if he made any further statements critical of GM food, he doesn't have a background in science himself. James Wilsden, head of science at the Demos think tank, described Taverne as being 'about as useful to science as Robert Kilroy-Silk is to race relations'.

Sense About Science director, Ellen Raphael, said: 'a little check-ing goes a long way.' This same organization claimed, in response to concerns raised by various celebrities, that 'most [chemicals] leave quickly but some stay: asbestos and silica in our lungs, dioxins in our blood. Do they matter? No!'

Another 'expert' declared that if cancer is increasing, 'it's because people are living longer'. This is hard to substantiate for all kinds of reasons, not least the fact that according to the US National Cancer Institute, childhood cancers have been increasing by 1 per cent every year since the fifties.

If ever there were a need for fact checking. . .

At the height of the public backlash against GM food, when the political and scientific establishments appeared to be losing the PR battle, 114 scientists wrote to the then prime minister, Tony Blair. 'Genetic engineering of plants', they complained, 'has been reduced to a matter of consumer pref-erence.'

For these scientists, the opinion of mere consumers appar-ently counts for nothing. Nor, presumably, does their right in a democratic system to make choices. This is the arrogant side of science; a belief that it should have a monopoly over the debate. We've seen it in the field of GM food, and we've seen it again in another, newer area: nanotechnology.

For enthusiasts, nanotechnology is about as exciting as science gets. For others, it's nothing short of terrifying. When I interviewed Bill Joy, founder and former chief scientist of technology giant Sun Microsystems some years ago, he told me, 'I think it's no exaggeration to say that we are on the cusp of an extreme evil, an evil whose possibility spreads well beyond weapons of mass destruction.'

It's not hard to see why nanotechnology is causing such a mixture of excitement and fear. The ability to manipulate materials at the level of the nanometre (one billionth of a metre) makes it possible to reorganize the very structure of substances at a scale where they no longer behave in the same way that they do in larger forms. It makes it possible to engineer new materials with properties never before experienced in nature.

Virtually everyone, enthusiast and critic alike, accepts that the upsides of nanotechnology – super-medicines, wear-proof car tyres, portable water purifiers and so on – are great. But many believe they are at least matched by downsides. In my interview with Bill Joy, he described the nightmare scenario painted by the famous physicist, Eric Drexler, where '"Plants" with "leaves" no more efficient than today's solar cells might out-compete real plants, crowding the biosphere with an inedible foliage. Tough omnivorous "bacteria" could out-compete real bacteria; they could spread like blowing pollen, replicate swiftly, and reduce the biosphere to dust in a matter of days.'

Until recently, even though nearly £1.8 billion is invested each year in a technology that makes GM appear backward, the vast majority of people had never even heard of it.

This began to change when a critical report by the action group on Erosion, Technology and Concentration (ETC), was sent by the office of the Prince of Wales to a national newspaper. But instead of welcoming a public discussion, the experts, and even the then science minister, Lord Sainsbury, bitterly accused the prince of meddling. He was using his position, they said, to distort the debate. The problem was that until that point, there *was* no debate.

Despite the gold-rush excitement fuelling its phenomenal expansion, there were no signs that the government or the nanotech industry were remotely interested in starting one.

On the contrary, the nanotech industry was intent on avoiding the mistakes of the GM industry, which attempted to win over sceptical consumers with an expensive and disastrous public relations campaign. The nanotech industry understood that consumer engagement translates to consumer demands: precaution, control and the right to choose.

So it was a reluctant industry that was dragged into the spotlight following coverage of the ETC report, and a still more reluctant industry that was then subjected to a government-initiated year-long investigation by the Royal Society and Royal Academy of Engineering.

When the results were published in 2004, many of the ETC group's concerns were vindicated. Whereas the regulators had maintained that nanotechnology was adequately covered by existing regulations, the report recognized that chemical substances at the nano scale can behave differently to the same substances in larger forms.

But the report wasn't exhaustive. 'It didn't consider the risks inherent in a merger between nanotech and biotech,' commented Jim Thomas of the ETC group. 'If you consider GM and nanotech as separate spheres of science, you can then dismiss self-replication as an irrelevant concern. But the fact is that nanotech and biotech are already converging to create hybrid materials, machines and living organisms.'

While not going as far as ETC might have wanted, the review was a start, and following its release, consumers can expect at the very least minimal protection by the regulatory system. That is in spite of an irresponsible former science minister, and because of the thankless and tireless work of groups like ETC that are doing the job he was paid to do himself.

As science reaches into new realms, what is clearer now than at any other time is that regulation of scientific progress is too important an issue to be left to corporate-funded scientists. Without a robust approach, it is inevitable that the market will race ahead of the science itself, and consumers will simply be left behind to deal with the mess. Certainly, that's been the experience with chemicals.

Chemical-Induced Puberty

Dr Marcia Herman-Giddens is a long-standing professor of child health at the University of North Carolina School of Public Health. In the years leading up to 1997, she began to notice a change in her young patients. 'I was seeing a lot of young girls coming in with pubic hair and breast development,' she said, 'and it seemed like there were too many, too young.'

So she embarked on what is still the largest ever study of sexual development in American children to date, in which 225 physicians examined more than 17,000 young girls. The results, published in the academic journal *Pediatrics*, were shocking even to the report's authors.

At the time, the standard medical line was that signs of puberty, including breast development and pubic hair, could be seen in 1 per cent of girls under the age of 8. By the end of the survey, that figure had risen to an astonishing 27.2 per cent for black girls, and 6.7 per cent for white girls. Worse, the doctors reported that 1 per cent of the 3-year-old girls they examined had swollen breasts or pubic hair.

Something in the American system is literally robbing children of their childhood. But what is almost more disturbing than the statistics themselves is the near blanket refusal on the part of the American political establishment to address the problem. Indeed the campaign for reversing, let alone understanding, these appalling trends has scarcely begun. How is that possible?

There is a solution, and we know what it involves. For years scientists have documented alarming changes to the development of sexual organs in countless species. The majority of lakes in the state of Florida, for instance, contain alligators unable to breed because their testes have been feminized to resemble ovaries. In one region of Montana, where 254 male deer were removed after road accidents, 67 per cent of the carcasses had severe genital abnormalities.

There is an abundance of examples that reveals a general problem. And at its root, according to all available science, is a wide family of endocrine-disrupting chemicals that to varying degrees mimic the effects of oestrogen.

Numerous studies have confirmed the link. Dr Walter Rogan, the former clinical director of the US National Institute of Environmental Health Sciences, for instance, measured the levels of PCBs and DDE in a group of pregnant women with a view to monitoring their children after birth. The girls with the highest exposure to the chemicals entered puberty eleven months earlier than their contemporaries. His study led to three separate tests on rats and mice, involving 'oestrogenic' chemicals like DDT, methoxychlor and PCBs. All produced results consistent with his findings.

And the phenomenon is not by any means restricted to the US. A study in Puerto Rico has shown that girls with early breast development have disproportionately high blood levels of phthalates, chemicals used in cosmetics, toys and plastic food containers.

Even the normally sluggish World Health Organization (WHO) has cautioned that 'the biological plausibility of possible damage to reproductive and developing systems from exposure to endocrine-disrupting chemicals is strong.'

Individual chemicals are occasionally banned. The pesticide atrazine, for example, which causes male frogs to grow ovaries in their testes at levels thirty times lower than those set by the US Environmental Protection Agency, was banned in 2004 by the EU after many years of campaigning.

The science is clear, and the effects are truly appalling. But despite this, there is no crisis committee, no urgency, no action to protect our children. Is 'precocious puberty' just another trend, like rising cancer, asthma and allergies, that we will simply have to get used to? Will puberty-slowing drugs become the new Ritalin?

Roughly 400 million tonnes of chemicals are produced in tens of thousands of varieties every year. Most have never been tested for their effects on health and the environment. In the last twelve years just 140 chemicals have been subjected to detailed risk assessment.

According to Vyvyan Howard, a senior lecturer at the University of Liverpool, there are currently 100,000 man-made chemicals in use, with another 1,000 added each year. The majority of these chemicals have been foisted upon the world without having been tested by the regulators, and those that have, have been analysed individually, removed from the context of the natural world in which they are being applied. What's more, it is known that these individual chemicals react synergistically when mixed with other chemicals, often rendering even the more benign chemicals lethal.

Testing these chemicals, even individually, would be a daunting task, but the idea of testing them thoroughly, and therefore in various combinations, would be impossible. 'To test just the 1,000 commonest toxic chemicals in unique combinations of three would', according to Howard, 'require at least 166 million different experiments – and this disregards the need to study varying doses. Even if each experiment took just one hour to complete and a hundred laboratories worked round the clock seven days a week, testing all possible three-way combinations of 1,000 chemicals, it would still take over 180 years to complete.'

Clearly that is impossible. But in the absence of proper

tests, it is equally impossible for the regulators to know that any one of these new chemicals is safe, as they often pretend. The UN Environment Programme said recently that 'Although the ecological impacts of chemicals are complex, some effects are well documented – the effects on various animals, birds and fish, include birth defects, cancers, and damage to nervous, reproductive and immune systems.'

Random tests on everything from raindrops to house dust, from blood to breast milk have revealed massive contamination by chemicals. Tests even on unborn children have revealed up to a hundred different chemicals. We are surrounded, and it cannot come as a surprise that we are seeing a huge increase in many varieties of cancer. It's a horrifying fact, for instance, that since the fifties, the incidence of testicular cancer has increased by 400 per cent.

When politicians boast about the robustness of our regulatory system, and the protection it offers us, we should consider that every one of the chemicals now held responsible for disrupting the sexual development of children was at one stage or another allowed onto the market by the regulators.

We need to move to a position where new products are assumed to be guilty until they have been proven innocent. Otherwise, when dangers are detected after a product has been unleashed, too many people with too much to lose will block honest progress. The precautionary principle is not the 'cowardice of a pampered society'. It is the only possible mechanism for avoiding potentially catastrophic errors.

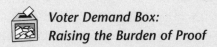

Voter Demand Box:
Raising the Burden of Proof

Our government must have, at the core of its regulatory approach, complete adherence to the precautionary principle where chemicals and new technologies such as GM are assumed to be guilty until they are proven to be innocent. We need to know that proper, independent and thorough research has been carried out – not merely presented by vested interest to our hopelessly compromised regulators.

Traditional Health

Progress has brought comforts and advantages for a great many people. But it has also been accompanied by the rise of some illnesses. Cancer, for instance, has not always affected one in three, or even two people, as is the case today. It has grown dramatically over the decades, despite billions of pounds and dollars invested in cancer research.

The problem is that much of that money has been spent on cures, and very little on trying to advance our understanding of the causes. Cancer has been treated by experts as an inevitability, something from which humanity can never escape. It has been seen as a disease of old age, and therefore almost a sign of our success as a species. But cancer is fast growing among the young, and in fact the earliest records relating to cancer reveal that it was very rare indeed among indigenous people.

For instance, in 1912 the Prudential Insurance Company of America published a report entitled *The Mortality From Cancer Throughout the World*. Its author was Fredrick L. Hoffman, chairman of the Committee on Statistics of the American Society for the Control of Cancer. Based on thousands of separate reports and all the available data, the author concluded that 'The rarity of cancer among native man suggests that the disease is primarily induced by the conditions and methods of living which typify our modern civilization.' He went on to say that 'The negative evidence is convincing that in the opinion of qualified medical observers, cancer is exceptionally rare among the primitive peoples.'

Around the same time, Sir Robert McCarrison, a surgeon in the Indian Health Service, observed 'A total absence of all diseases during the time I spent in the Hunza Valley [seven years]. . . During the period of my association with these peoples, I never saw a case of cancer.' Also writing of the Hunzas, Dr Allen E. Banik and Renée Taylor, in their 1960 book *Hunzaland*, described 'Their freedom from a variety of diseases and physical ailments. Cancer, heart attacks, muscular complaints, and many of the common childhood diseases. . . are unknown among them.'

Around the world, the story appears to have been the same. The Nobel laureate Albert Schweitzer wrote 'On my arrival in Gabon, in 1913, I was astonished to encounter no case of cancer.'

Dr Puzin, meanwhile, having studied 10,000 women in Senegal in the early nineteenth century, was only able to detect one case of breast cancer. Had Dr Puzin selected a random 10,000 women in the United States, he could have expected to diagnose at least 1,000 breast cancer victims.

Can the National Health Service Cope?

When the National Health Service was established in 1948, it was given a budget of just £170 million. One of its architects, Sir William Beveridge, explained at the time that no allowance had been made for an increased budget in the following decades, because a proper health service would eliminate illness. 'No change is made in this figure as from 1945 to 1965,' he said, 'it being assumed that there will actually be some development of the service, and as a consequence of this development a reduction in the number of cases requiring it.' The NHS now costs £100 billion per year.

Step Four

Food Quality, Food Security

A country which cannot afford art or agriculture is a country in which one cannot afford to live.
John Maynard Keynes

Considering that every voter has to eat, it's amazing that food security isn't one of the highest priorities for governments. Yet ours has repeated time and again that such a policy is unnecessary: Britain is food secure, we're told, because we can afford to buy produce from other countries.

In 2006, the then Secretary of State for the Department for Environment, Food and Rural Affairs (DEFRA), Margaret Beckett, explained the government's position:

> We do not take the view that food security is synonymous with self-sufficiency. . . It is freer trade in agriculture that enables countries to source food from the global market in the event of climatic disaster in a particular part of the world. . . it is trade liberalization which will bring the prosperity and economic interdependency that underpins genuine long term global security.

Even as food prices began to soar in 2008 on the back of rising fuel costs, declining harvests and increased use of land to grow biofuels, the government refused to budge, with DEFRA claiming that 'Because the UK is a developed economy, we are able to access the food we need on the global market. . . Climate change', it added, 'is likely to bring new challenges for the food security, not of rich countries like the UK, but of less developed, tropical, regions.' At the time, some fourteen countries faced food-related unrest, from 'tortilla riots' in Mexico to protests over the price of pasta in Italy.

Instead of ensuring a viable food base in the UK, the guiding policy is to leave farmers at the mercy of the global economy. The message has been consistent: farms need to expand and be internationally competitive, or get out of the way. Just like any other sector.

The trouble is, they can't be internationally competitive, and the vast majority will therefore have to do just that – get out of the way.

Over the years, British consumers have rightly demanded and won high standards in areas like animal welfare, but because the countries we buy food from face much lower hurdles, our own farmers have been put at a huge disadvantage. That, combined with high land value and labour costs, heavy regulation and an increasingly powerful retail sector, as well as the painful legacy of both BSE and foot and mouth disease, have all contributed to the farming industry's difficulties.

Only the very largest farms enjoy any security, with the smaller, more diverse, family farms teetering on the edge of viability. Farm incomes have been falling – down by 11 per cent since 1997 in real terms – and at various times in recent years, we have lost up to 1,000 farms a week. In short, we are seeing the near collapse of farming in the UK. And the consequences of that could be grave.

Until the 1830s, Britain was virtually 100 per cent self-sufficient. Today we import anywhere between 40 and 51 per cent of the food we need. In the last ten years alone, the area put down for vegetables has declined by 25 per cent, while 90 per cent of all fruit is imported. We are becoming a nation increasingly dependent for our most basic survival on other countries.

But how long can this be sustained? Like all global systems, the global food economy is vulnerable to international political crises, rising fuel costs, terrorism, population increases and any number of other factors. Not only that, but it also heaps increasing pressure on the world's environment. That the government continues to ignore these pitfalls suggests it isn't paying much attention.

According to a report by the UN, a combination of drought, deforestation, industrial agriculture and climatic volatility is responsible for the loss of 250 million acres of fertile soil each year, undermining the food security of 1.2 *billion* people in 110 countries. Australia, a significant originator of our food, recently experienced its worst drought in history, which reduced its wheat crop by more than 50 per cent, and China's grain harvest has also fallen – by nearly 10 per cent over the last seven years, leaving it increasingly reliant on world markets to make up the shortfall. This at a time when world grain stocks are at their lowest level in more than thirty years.

And if this sounds bleak, the future hardly seems rosier. It is estimated that if just one sixth of the West Antarctic ice sheet melts, the resulting one metre sea-level rise will cause 30 per cent of the world's total cropland to be flooded.

So why *are* we pursuing policies that render us so vulnerable? Even while the importance of food security in the UK has risen, actual food security is being eroded. We are becoming reliant for our survival on a fundamentally unreliable system. We are effectively banking *everything* on the assumption that we will always be able to pay for the food we need, that the world's breadbaskets will always be able to provide it, that cheap oil will always be available to distribute it.

Politicians are notorious for their short memories and still shorter-term perspectives. But on this issue of food security, successive governments are guilty of truly breathtaking naivety.

The battle of ideas

For decades, the dominant view has been that feeding the world requires ever bigger, more industrialized farms. But that is beginning to change as we are witnessing the mass erosion of the world's great breadbaskets, like the Punjab, which is more than half eroded, and the collapse of water tables all over the world.

Industrialization of agriculture has sometimes increased yields, but it has also destroyed the very land we depend on, and displaced hundreds of millions of farmers. For some, the shift has undoubtedly brought benefits. But for others, it has been a disaster. The UN estimates that nearly a billion people exist in urban squalor – well below the poverty line.

In 2008 the UN Food and Agriculture Organization and the World Bank issued a shock report that called for radical change. It was put together by 400 experts, and came at a time when surging food prices were fanning violent protests in Haiti, Egypt and the Philippines, and exposed serious concerns about the global food supply in coming decades.

'The status quo today is no longer an option,' it said. 'We must develop agriculture that is less dependent on fossil fuels, favours the use of locally available resources and explores use of natural processes like crop rotation and use of organic fertilizers.'

71

As a geopolitical statement there can be few more startling about-turns. Both organizations, it appeared, were wedded to the bigger, industrialized farming models, yet here they were advocating a system that seemed to go against all their instincts. But while the message might be surprising, what they have discovered is by no means new or revolutionary. Small-scale diverse farms may be less productive per unit of labour (they are more labour intensive) but they have long been proven to be more productive per unit of land.

Studies conducted in a variety of countries around the world have consistently shown that smaller, more diverse farms are in fact more productive than larger intensive farms – and not just by a little. One study by the Food and Development Policy Institute in the USA suggested that small traditional farms are 200 to 1,000 per cent more productive than bigger farms. In Brazil, former minister for the environment, José Lutzenberger, claimed that indigenous farmers produced at least three times more food than intensive farms.

In 2000, the UN researched over 200 food-growing projects in the southern hemisphere and found organic techniques increased yields by anything between 46 per cent and 150 per cent. A 2007 review of this same data found that organic production methods increased yields in subsistence agriculture by 80 per cent. It concluded that if world agriculture went organic, overall food production would increase.

Another UN study of 114 projects in twenty-four African countries found that yields had more than doubled with the introduction of organic practices. The same study found that as well as increasing yields in these countries, organic farming brought other benefits, for instance increased soil fertility, water retention and biodiversity.

The choice now is between the status quo and a shift to a more localized, diverse system of farming. But while many people now recognize the short-sightedness of the former, not least because of its appalling effects on some of the world's great breadbaskets, the political establishment is still determined to find ways to extend the intensive monoculture model. This is where the hugely controversial GM comes in.

Intensive agriculture and GM

At one point, in the late nineties, GM appeared to have been crushed by popular opinion following an unprecedented consumer backlash. At the time, a MORI poll showed that two thirds of people didn't want to eat any GM food. The share price of the leading biotech firm Monsanto plummeted by 40 per cent.

The big supermarkets were quick to remove GM from their shelves. Iceland was the first in 1998; within months practically all the UK's leading supermarkets bowed to consumer pressure and cleared their shelves of GM food. In July 1998, a statement from the House of Commons Refreshment Department confirmed that MPs were being served food that avoided GM ingredients 'in response to the general unease about such foods expressed by significant numbers of our customers'.

The industry was enraged, but its reaction provoked still more revulsion among consumers. Scientists complained to the prime minister, en masse, that 'Genetic engineering of plants has been reduced to a matter of consumer preference.' The implication, clearly, was that it should be forced upon the ignorant masses. But if one thing separates democracies from authoritarian regimes it is the ability of the people to say 'no' to unwanted products.

GM firms, like Monsanto, were caught infiltrating 'independent' scientific committees. One biotech consultancy group, Promar International, succinctly encapsulated the industry's hopes that 'Over time the market is so flooded that there's nothing you can do about it, you just sort of surrender.'

The truth is, there is no natural market for GM food. People do not choose to buy GM food, and are unlikely to do so. So the only remaining market is a captive one where people have no choice. If GM foods are unlabelled, then people cannot discriminate. And those that are desperate have even less choice. Former US President George W. Bush made pharmaceutical aid in the Third World conditional upon the acceptance of GM food.

The British government has always been wildly pro GM food. But it has had to navigate its way through continued hostility. In 2003 it launched a GM 'dialogue' comprising three strands. The first concerned public opinion. The results were overwhelmingly hostile. Strand two was a GM science review, which acknowledged that 'many of the uncertainties and gaps in knowledge addressed, for example in long-term impacts on health or the environment and the coexistence of GM crops with other crops, coincide with concerns expressed during the public debate'.

Finally, a cost-benefit analysis by the government's Strategy Unit found that any economic benefit of GM crops was likely to be limited, at least in the short term, and outweighed by other developments.

Yet despite years of disappointment, the UK government continues to back agricultural biotechnology – to the tune of £50 million a year, compared with just £2 million for straight organic research. Why?

Faced with enormous problems – population growth, food shortages, erosion and water depletion – it's far easier for decision-makers to pin their hopes on a future technology than it is to tackle these problems systemically. It's easier to imagine a magic bullet than to consider a wholesale shift in the global food system, even if that is what is required.

GM is a promise. Almost every hope that is pinned on the technology is still just that: a hope. Indeed, a problem doesn't exist that the GM firms don't claim an answer for. In 2003, *The Times* ran a headline that read: 'Plants may warn of bio-terror'. From pollution-eating ocean microbes to plants capable of growing without water, the GM industry always seems to have an answer.

Undoubtedly, this has suited politicians. Above all, GM is a technology that fits the same agricultural model that they don't want to have to challenge. And it offers a way out of the difficulties we now face without the need for any radical shift in policy. But the fact still remains that, beyond the hype, GM has simply not delivered on its promises.

Claims that genetic engineering would produce new drought- and salt-tolerant varieties were made twenty-five years ago. By 2005, the US Department of Agriculture had reportedly seen over 1,000 applications to test GM plants for drought tolerance. None appears to have been successful. Nor are there any GM crops producing a new generation of cheap pharmaceuticals.

Since the first GM soya crop of 1996, only two GM traits have been commercialized anywhere in the world: herbicide tolerance and insect resistance. Most of the plant varieties created by GM firms are designed to be resistant to chemicals sold by the same companies.

So it's not about increased yield. It's about developing new vehicles for selling the same old chemicals. Ultimately it's about control of the food chain. Indeed, none of the existing GM crops in commercial cultivation is engineered specifically for yield increases. When pressed, the former head of GM giant Novartis admitted that 'If anyone tells you that GM is

75

going to feed the world, tell them that it is not.'

Nor have GM crops reduced the need for chemicals in agriculture. Unintended breeding between different GM varieties in North America has created 'superweeds' so virulent that more chemicals are needed to tackle them. One Canadian government study found superweeds at every site it examined. Even the traditionally pro-GM US Department of Agriculture has revealed that over three years there was a net increase of 73 million pounds of pesticides on land planted to GM crops. Partly as a result, the same organization has publicly questioned the take-up by farmers of GM crops given their 'mixed or even negative' financial impacts. The safety issues surrounding GM are still more worrying (see Step Three).

We are what we eat

It is time for us to decide in which direction we would like to travel. Britain faces a mounting public health crisis, largely caused by poor diet. It's estimated that a third of all deaths from heart disease and a quarter of all cancer-related deaths can be attributed to bad eating habits. The UK now has the highest rate of obesity in Europe, with one in four of us considered overweight. If trends continue, half of all children will be obese or overweight by 2020. The cost to the NHS of treating patients for such diseases is already £1 billion a year.

But it is not just our physical well-being that can be affected by eating badly. A growing body of research demonstrates the impact of diet on mental health, and there is also growing evidence linking diet to anti-social and criminal behaviour.

This was made clear during an experiment at HM Young Offenders Institute in Aylesbury. When prisoners were given nutritional supplements, behavioural problems were reduced by a third. Although this evidence has been recognized by the

World Health Organization, it has yet to be recognized by the UK Home Office. Meanwhile, the Dutch Ministry of Justice has estimated that addressing nutritional deficiencies is potentially so cost effective that it would allow them to improve services while achieving an estimated 18 per cent cost saving.

Over recent years, food has become both a social and a political issue. Sales of organic food have boomed, as have the number of farmers' markets. Celebrity chefs urge us to eat well, to monitor where our food comes from, to buy local and organic. This groundswell in feeling shows that there is an appetite for a major policy shift.

There are some levers available to government that would trigger immediate change – at no extra cost to taxpayers. For example, the government spends approximately £130 billion of our money each year on goods and services. £1.8 billion of this is spent on food for schools, hospitals and prisons. Every year the NHS alone buys over 300 million meals, spending £500 million in the process. Currently only around 2 per cent of government-procured food is sourced locally. That needs to change.

If this money was invested wisely it could provide a massive boost to farmers and would make good quality food available to a large number of people. If just 20 per cent of the money spent on food in London's sixty-nine hospital trusts was spent on local food, that would provide a boost for local farming and food businesses in the south-east of over £3 million a year. If 3,600 primary and secondary schools sourced 50 per cent local and 30 per cent organic produce, it would create a new market for local and organic produce worth £66 million.

Investing in this way would provide nutritious food to some of the most vulnerable members of our society – children, through the school meal service, patients in hospitals, the elderly in care homes. It would benefit the environment through reduced 'food miles', reduced congestion from freight, and by encouraging farming with fewer pesticides, and ultimately, greater biodiversity.

And it doesn't have to cost more. The Royal Cornwall Hospital Trusts, for example, already source 83 per cent of their food from local Cornish farmers. The programme has cut food miles by 67 per cent, and even more impressively its dinners have been rated 'very good' or 'excellent' by 92 per cent of patients. All that, and the overall cost has remained the same.

The story of the Merton Parents (in the next box) shows what is possible and why feeding children with good food works. But it's not just eating food. Children need to know about food, where it comes from, how it is produced. Every school should include food-growing in the curriculum. For some that will mean twinning with farms. For others it will mean literally building their own smallholdings.

Growing food – as a process – has a value in itself. It's hard to quantify, but it is nevertheless real. Catherine Sneed, a young counsellor in San Francisco's County Jail, noticed early on in her career that the same people kept on returning to jail. Inspired by *The Grapes of Wrath*, a novel in which connectedness to the land binds families together, she set up a small prison garden. Inmates loved it, and the project flourished. The food they grow feeds hundreds of low-income families in the area, and inmates who take part in the project are a staggering 25 per cent less likely to return to jail than those who don't.

If growing food is therapeutic for California's prisoners, there is every reason to believe it will be good for our children too.

School Dinners: A Success Story

In 2005, a group of parents in Merton came together to improve the quality of school food. They called themselves Merton Parents for Better Food in School.

They set themselves a few key goals: to win funding for a working kitchen in every school; to improve the quality of ingredients and cooking standards; to encourage schools to improve lunchtime arrangements and to persuade the local authority to sign up to a good quality school meal service. It was an ambitious list, and no one knew if it would work.

But it did. Merton's parents convinced the council to put aside £450,000 to refurbish primary school kitchens and allow them to produce fresh food on site. They also set up a farm-twinning scheme with a nearby farm. The council has agreed to manage the scheme, which will eventually involve every school in the borough.

Jackie Schneider, chair of the Merton Parents, said 'It is amazing what can be achieved when the whole community pulls together. It was the combination of parents, governors, catering staff, schools and local government working together that finally got thirty-nine kitchens built in Merton primary schools and a new improved menu.'

 Voter Demand Box

Better food for children, better income for farmers
The £2 billion annual food budget for schools, hospitals, military barracks and prisons should be invested as a matter of course in the most sustainable, most local produce.

Food as part of the curriculum
Every school should be helped to include food-growing in the curriculum. For some that will mean twinning with farms. For others it will mean literally building their own smallholdings.

Reforming the food subsidies

In recent years there has been a welcome shift from the old EU system of handing farmers subsidies for producing as much food for export as possible, towards paying farmers who manage the environment responsibly. But it has been a sluggish process and there is much more that needs to be addressed. Paying farmers to grow food is clearly wrong – after all, that's their job. But where farmers are looking after land in such a way as to benefit society, without being rewarded by the market, there is a strong case for subsidies.

For example, the Common Agricultural Policy (CAP) – the system that administers the EU's programme of subsidies – should compensate farmers, reward them even, for not over-draining their land. One of the reasons flooding has become so severe in the UK is that for decades we paid farmers to bring marginal land into production. That meant draining wetlands, clearing wooded slopes, straightening rivers and embanking water meadows. Up to 40 per cent of the Severn River catchment – site of the worst floods in the wet summer of 2007– was probably degraded in this way. Today, farmers need to be encouraged to farm with water in mind. They also need compensation for maintaining traditional landscapes. Upland hill farms, for instance, will never be economically viable.

There will always be a need for a fixed sum of CAP's

budget to be directed towards paying farmers for such services, and this needs to be refocused on the areas of greatest need.

The remainder should then be invested in renewing the infrastructure needed to support a local food economy. Any infrastructure for local food that once may have existed has all but disappeared with the advent of the centralized supermarket distribution and supply systems. Abattoirs have shut down, processors and small independent food shops such as butchers, greengrocers, village shops and convenience stores are on the decline. A thriving local food economy needs this infrastructure, and without it a local food renaissance can only remain an aspiration.

On top of reforms to the way our taxpayer subsidies are spent, we also need to address the less obvious indirect subsidies used to support the wrong type of food production. Professor Jules Pretty of the University of Essex has calculated that if all UK farming was organic, the external costs of agriculture would be reduced by £1.6 billion. An example of this is pesticide use. Three quarters of UK drinking water sources are contaminated with pesticides above the accepted levels. Removing them from the water supply in the UK costs taxpayers about £300 million annually. BSE, which was the result of non-organic farmers cutting costs by feeding cows an inappropriate diet, is believed to have cost £4.5 billion.

None of these costs appears on the price tag. Pesticide manufacturers pass on the costs of cleaning up pesticides to farmers, who in turn pass it on to water companies, who in turn pass it on to us consumers via water bills. Polluters get a hidden subsidy from anyone who pays a water bill. The non-polluter – the organic farmer – receives no such subsidy.

So when we talk of 'cheap food' it is worth remembering that more often than not, we have paid twice for it.

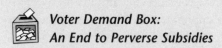

Voter Demand Box:
An End to Perverse Subsidies

Subsidies in the public interest
We should insist that the bulk of our CAP funds be used exclusively to reward farmers for managing their land in a way that is beneficial to society, but unrewarded by the market. The overall CAP budget should shrink over time, with the diminishing part of the budget being used to invest in rebuilding the small scale domestic food infrastructure we've lost over the decades.

Make the polluter pay
The government needs to set about identifying and calculating the indirect subsidies to agriculture. The costs of inappropriate or polluting agriculture needs to be taken up by the polluter – not the taxpayer.

Breaking the supermarket strangle-hold

If we are to develop a truly sustainable food economy, we need a diverse retail sector. Given that the four largest supermarkets now control 75 per cent of the retail market, we're a long way off.

There is no doubt that the big supermarkets have increased the range and the sheer variety of foodstuffs on offer. But beyond the sliding doors of our vast superstores, the price is being paid. The growing dominance of the multiple retailers has taken its toll – on farmers, on small shopkeepers and on the environment. More than 2,000 small independent shops close down every year. Between 1986 and 1996, the UK lost half its small grocers. Towns and villages have lost an important part of their community, sacrificed by supermarket discounts and the dispersal of local trade.

Some would argue that this is a small price to pay for the jobs and prosperity that out-of-town supermarkets and retail parks create. But the figures don't stack up. According to the New Economics Foundation, money spent on locally produced food generates almost twice as much income for the local economy as the same amount spent in a typical supermarket.

A government report on superstores found that 'There is strong evidence that new food superstores have, on average, a negative net effect on retail employment.' Supermarket power is also felt by food producers, whose own bargaining power with the big chains is now so negligible that they effectively have to accede to their terms. The over-dominance of supermarket power represents an effective monopoly, and it needs to be challenged. Local referendums, recommended in Step Two would help. But we need more.

In the UK, the All-Party Parliamentary Small Shops Group is calling for a moratorium on further supermarket mergers, a new retail regulator and more power for local authorities in deciding on planning applications. It's a good start but it doesn't go far enough to redress the imbalance caused by the might of the chains.

'The Great Food Swap'

'Taking coals to Newcastle' is an old English phrase, used to describe something utterly pointless. Perhaps in the future we will find a food reference to make the same point. Consider this: in 2007, Britain imported 46,085,000 tonnes of poultry meat while exporting 29,358,000 tonnes. Britain also imported 50 million litres of liquid milk when it was exporting 539 million litres. The UK exported   87,000 tonnes of pork to the EU, while importing 461,000 tonnes.

Currently, planning is biased in favour of the development of the superstore over alternative retail outlets. The planning system is supposed to prioritize the protection and enhancement of town centres and local neighbourhood shopping centres. But in reality, the government is planning to relax these laws to encourage more out-of-town supermarkets. One supermarket boss recently admitted to me that it is easier than ever to gain permission to build out-of-town developments.

On behalf of our farmers, we should demand that our supermarkets adopt a formal code of practice. There is already a voluntary code, but it is weak and ambiguous, particularly in relation to terms such as 'reasonable behaviour'. A clear definition would make it easier to prove a breach of the code. What is clear is that the current imbalance of power makes healthy competition impossible.

Voter Demand Box:
Break the Armlock

Remove the planning bias
At every level, planning policy should be geared towards town-centre enhancement, 'walkability' and maintaining the viability of independent shops.

A code of practice

Supermarkets with more than 8 per cent market share need to sign up to a new, strengthened and mandatory code of conduct. Amongst other things, this would ensure fair practice when they deal with farmers.

Moratorium on mergers

It is already a fact that the big supermarkets enjoy disproportionate market share. Parliament needs to agree a reasonable cap, and set a timetable for achieving it. That may mean breaking up the largest monopolies.

An impossible international competition

When faced with the startling facts of the decline in British agriculture government farming experts tend to use them as fuel for the argument that the only way forward is through competition. Instead of looking for new ways to help farmers and communities become both viable and stable, they suggest this decline proves that farmers need to consolidate, grow and become more efficient if they are to cope with international competition.

But this premise is fundamentally wrong. It has become virtually impossible for British farming to contend with competition from abroad. Britain's farmers compete on a horribly uneven playing field.

For one thing, our approach to regulating farming is hugely intrusive, and has at its heart the assumption that farmers are either incompetent or dishonest. Every stage of every process is necessarily 'policed' by government. The sheer volume of red tape, laws, directives and guidelines imposed on farmers means they are forever hampered. The effect, perversely, is that smaller operators cannot afford to keep up with the rising cost of production. The equipment required by ever-changing laws, the new milk-bottling machines, the hygienic tiles, all make small-scale production uneconomic.

It's not uncommon for small and family farmers to spend a quarter of their income employing secretaries to help them navigate the bureaucracy. But who in government knows more about farming than the farmers themselves?

In addition to this, the standards required of our farmers are already higher than most, and as a result they suffer from intense competition with cheaper imports from countries where standards are less demanding. The effect, perversely, is that we merely end up undermining our own producers and importing the same low-standard food from elsewhere because it is cheaper. It's an unfair competition, and for the sake of all of us, not just our farming industry, it cannot be allowed to continue.

So what choice is there? Obviously, we could go back to the way things were and lower our standards to match those of the importers, turn our back on progress and allow the kind of practices consumers abhor. Or we could be brave and demand that the produce we import meets the same standards that are expected at home.

But despite its logic and common sense, this kind of policy is specifically outlawed by both the EU and the World Trade Organization. It would take a strong government indeed to stand up to these international trade rules, but Britain would not be alone. While our politicians prevaricate and hide behind international trade rules, France has had the courage to demand the same animal welfare standards on meat imports as those required within her own borders. Their stand proves that where the law makes common sense a crime it is the law that should yield, not common sense.

Raising standards at home

Our standards may be higher than most, but there are profound problems in the way we raise some of our livestock. High-profile television chefs have recently put the spotlight on, for example, battery chickens, and many viewers have been shocked that they have, unknowingly, been supporting practices they would never condone.

The conditions in which many of our animals are kept are entirely unnatural, and it is only because of our discovery of antibiotics in the first part of the twentieth century that we are able to maintain them at all. However, there is a cost, and it goes well beyond issues of cruelty.

When they were first discovered, antibiotics were truly the miracle cure our species had longed for. More than any other medical advance, they have saved millions of lives. But today most antibiotics aren't used for saving lives.

Farmers discovered that by using antibiotics on their animals, they could massively increase their stock, and keep them in conditions that would normally have killed them. Animals could be brought in from the fields and cramped together on concrete floors, in airless rooms. Breeds could be developed that produce unnaturally high yields – chickens, for instance, that fatten in just seven weeks, instead of twelve.

But the routine use of antibiotics to accelerate growth and to keep animals alive in appalling conditions has led to a rapid spread of antibiotic resistance among harmful bacteria.

We have known for some time that overuse of antibiotics on farms leads to antibiotic resistance in food poisons such as salmonella, and there is now compelling evidence linking the same practice with the rise of lethal new strains of MRSA and E. coli. Already, a quarter of all MRSA cases in Dutch hospitals involve the animal strain.

The process is no mystery. When bacteria are exposed to antibiotics, many die, but some genetically mutate and develop resistance. As the drugs become stronger, so too does the resistance, until eventually superbugs emerge that are virtually unstoppable. In 2007, an average of twenty-three people died every day in Britain as a result of two varieties of superbugs.

Although we know this, pressure from agribusiness has prevented successive governments from acting to phase out the practice. The economic impact on factory farms, they say, would be heavy. But against the costs involved in battling increasingly deadly superbugs, those costs are minor. We are literally squandering the most important medical advance of all time, and all for short-term profit.

Today's Best Standards

The Soil Association promotes the world's strictest organic certification system. It covers a wide variety of concerns such as use of natural soil biology rather then pesticides and fertilizers, rotation, nutrient recycling, animal welfare, whole-life-cycle costs of food production and distribution.

Another standard – Linking Environment and Farming (LEAF) – is an industry certification scheme that encourages efficient farming systems that look after the land and the rural community. LEAF is a middle ground between the baseline levels and the high standards of the Soil Association's organic certificate.

A new approach

American poet Wendell Berry wrote:

> Once plants and animals were raised together on the same farm they therefore neither produced unmanageable surpluses of manure to pollute the water supply, nor depended on such quantities of commercial fertilizer. The genius of American farm experts is well demonstrated here: they can take a solution and divide it neatly into two problems.

The industrial food system has fed more people than ever before, but fed many of them badly – and destroyed much of the environment in the process.

It has also brought huge vulnerability. In his book *The Killing of the Countryside*, Graham Harvey laments: 'The small, mixed farm was the nation's insurance against catastrophic change. The politicians have traded it for international competitiveness.'

A better government procurement policy would provide significant support for local sustainable agriculture. A more balanced retail sector would ensure fairer income for farmers. A reformed CAP would reward farmers for farming in environmentally beneficial ways. An even playing field would enable British farmers to adhere to high standards without being squeezed by international competitors.

The effect would be a more diverse, localized and natural food system, one that would nourish rather than exhaust its own base; in other words, a constant economy. That's the kind of food and farming most people actually want. Nothing could be more important for our health, the health of our children and the health of our landscapes and environment than the food we produce and how we eat it.

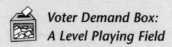

Voter Demand Box:
A Level Playing Field

Good cop/bad cop

As many environmental payments as possible should be based on existing farm assurance schemes. With standards imposed by these existing bodies, DEFRA and its agencies could outsource a good deal of inspection and regulation. It offers real incentives to farmers – payments for environmental outputs, status as an 'approved' farm, and less regulatory intervention. Where that trust is abused, the farmer would revert to a routine of regulation, and loss of assurance status, which might in turn lead to loss of contracts with retailers. This 'good cop/bad cop' approach would be welcomed by the vast majority of farmers.

Campaign for higher international standards

Both the World Trade Organization and the EU prevent national governments from favouring their own farmers or producers. This is wrong, and we can lobby for reform. It should be possible for nations to insist that imported foods meet their own standards. France already demands the same animal welfare standards of imported food as it expects at home. We must do the same. Our government must also work to raise standards internationally. This should involve an urgent review of global trade rules in the context of global food security.

A British label

We would all benefit if there were a minimum standard label for British food. Consumers would then be able to apply pressure on the large retailers to favour sustainable produce, which would then become cheaper through an altogether healthier economy of scale. And wherever possible, for instance in public procurement through schools, hospitals and prisons, support for the standard should be mandated. But the label must reflect the truth. At the moment, food can be labelled as 'British', no matter where it has been produced, as long as it is processed in Britain. This practice must be outlawed.

Phase out the antibiotics

A timetable needs to be set for the banning of routine antibiotic use on farms for growth promotion, and to 'prevent' disease. Use should only be allowed to treat sick animals. In addition, the government needs to lobby for an EU-wide ban, and to push the World Trade Organization to recognize the move as a non-trade related issue. In other words, countries would be free to ban imports of meat raised on a diet of antibiotics.

Protection against GM

We should follow the lead of the Welsh Assembly government which has introduced a law requiring farmers growing GM crops to publicly register with the government, and to assume full financial responsibility in the event that their GM crops cause any loss of earnings for neighbouring farms. An example would be if an organic farm loses its status. Under EU law, Wales is unable, and therefore unwilling to declare itself 'GM-free', but such a measure makes it highly unlikely that Welsh farmers will take the risk of growing GM crops.

Local Food, Global Poverty

Does pushing for more localized food production mean undermining poor farmers in the developing world? There is no reason it would have to. There will always be trade between countries in those goods which cannot be produced locally – coffee, cocoa, exotic fruits, etc.

However trade will be affected by factors other than policy decisions. The abundance of cheap oil, crucial for trade, is a luxury that we should not count on in years ahead. Concerns are also rising about the effects of soil erosion and water depletion on global food production capacity. The global food economy is vulnerable on multiple levels, and its demise will hit the poorest farmers hard.

For years, poor countries have been encouraged by international lending agencies and the IMF to adopt agricultural models geared towards export, often at the expense of their domestic food security. It is clear this cannot last. Rich nations should be supporting poor nations as they move away from uncertain agricultural export and towards more stable means of growing their economies.

Is it the case, as Ethiopian environment minister Tewolde Egziabher believes, that the export model creates vulnerability and food insecurity? Egziabher's view is echoed by Davison Budhoo, a former IMF economist who said recently that export orientation for poor nations 'has led to the devastation of traditional agriculture and the emergence of hordes of landless farmers in nearly every country in which the fund operates.' Localization, then, is not the problem for the poor: it is globalization they need to worry about.

Step Five

Save Our Seas

Wilderness is the bank on which all checks are drawn.
John Aspinall

In 1497, the explorer John Cabot described how his ship's progress had been hindered by the sheer volume of cod off the coast of Newfoundland. In dispatches returned to his sponsor, King Henry VII, he wrote that his men 'took so many fish that this kingdom will no longer have need of Iceland, from which country there is an immense trade in the fish.'

In the centuries that followed, ships travelled from around the world to share in the bounty described by Cabot. Right up until the fifties, Newfoundland appeared to offer a limitless supply of fish – the vast north-west Atlantic region off the coast of Labrador and to the east of Newfoundland yielding an annual average catch of about 250,000 tonnes of northern cod.

And so it could have continued. But in the post-war world, a new type of fishing emerged. Giant 'factory trawlers' began to appear, sometimes from countries thousands of miles away. The small craft favoured by traditional fishermen could not compete with much larger vessels and were soon rendered obsolete. For twenty years, the impact of these Spanish, German, British, even East Asian vessels appeared minimal.

With giant nets and on-board processing equipment, they were able to exploit the seas around the clock.

By 1968, the registered catch was 800,000 tonnes per year, yet by 1975 that had dropped to just 300,000 tonnes. Canada's politicians began to worry – and with good reason. Total allowable catches for northern cod had been introduced in 1973, but led to little in the way of fleet restrictions. In July 1977, Canada decided to introduce a 200-nautical-mile fisheries management zone to help depleted fish stocks. Foreign ships were banished and the catch more or less halved. But instead of allowing the stocks enough time to recover, the fishing industry grew impatient, and before long Canada's fishermen had been replaced by a home-grown industrial fishing fleet every bit as powerful as that which politicians had banned. In 1978, the cod harvest hit an all time low of 139,000 tonnes.

Money flowed into the construction of so-called 'draggers', each one as big as a football pitch, weighed down with chains that dredged along the ocean floor, destroying everything in their path. They trapped whole shoals of fish, including the young and the inedible, and disrupted the ocean ecosystem.

Within ten years of the new laws, the catch had been allowed to climb again to 250,000 tonnes, where it remained throughout the eighties. Local fishermen, those with smaller, more traditional boats, began warning that serious damage was being done. They urged government scientists to review their optimistic findings. By 1986 it was obvious to scientists too, and they recommended that the government cut the total allowable catch by 50 per cent immediately. Concerned about the job losses such a move would incur, the government flatly rejected the recommendations. Six years later, however, their short-sightedness would result in crisis. In 1992, the fishery collapsed.

The Canadian government was forced to issue a two-year moratorium on cod fishing in its 200-mile zone. For the first time in four centuries, there was to be zero cod fishing in Newfoundland – and the consequences are still being felt today. Over 40,000 people lost their livelihoods. Many settlements are now nothing but ghost towns, communities ripped apart by the effect of overfishing. Despite billions of dollars put aside to help struggling families, Newfoundland remains an economic disaster zone.

It's hard to imagine a clearer example of how not to take care of an economy than the Newfoundland tragedy. In just four years, from 1990 to 1994 the Newfoundland cod population crashed from an estimated 400,000 tonnes to just 1,700 tonnes. This in a fishery that had reliably yielded an average of 250,000 tonnes of cod each year. It was, and remains, as one senior fisheries bureaucrat described, 'a calamity of almost biblical proportions'. It is something we are destined to repeat unless we heed its warning.

A tragedy repeated

In only a few decades, we have brought the world's oceans to the brink of exhaustion. Between 70 and 80 per cent of the world's marine fish stocks are either fully exploited, over-exploited, depleted or recovering from depletion. Fifteen of the seventeen largest fisheries in the world are so heavily depleted that future catches cannot be guaranteed. In the North Sea many once-common species such as cod, skate and plaice are now overfished. In fact cod stocks are on the verge of commercial collapse and common skate is nearly extinct. The Indian Ocean and South East Asian seas had average declines of species of more than 50 per cent between 1970 and 2000.

There is virtually no area of the world's oceans, nor

individual species, that is insulated from our activities. Despite serving as breeding grounds for 85 per cent of commercial fish, a third of the world's mangroves have been destroyed since 1990. In South America, it is closer to half. Albatrosses have suffered huge losses, as have whales, sea turtles, walruses, seals and crustaceans, while a paper published in *Nature* suggested that we have lost 90 per cent of the world's big predatory fish such as tunas and sharks. Another study published in *Science* predicts that all the world's fisheries will collapse by 2048 if trends are allowed to continue.

More worrying is the fate of the coral reefs. Crucial as breeding grounds for deep-sea fish, and regarded as the ocean's rainforests, more than a quarter have been wiped out. Of the rest, just 5 per cent are considered to be pristine. Savage fishing techniques, including dynamite and trawling, have wrecked huge areas of coral reefs, but the biggest threat is climate change. The more carbon dioxide we release into the atmosphere, the more is absorbed into the oceans which are becoming increasingly acidic as a result. Corals, and indeed ocean creatures, now have to cope with a 30 per cent increase in acidity.

The destruction of our marine environment is more than an environmental issue. About 200 million people depend directly on the fishing industry. For more than a billion people, fish is their primary source of protein. The export value of the world fish trade is greater than the combined value of rice, coffee, sugar and tea exports. If the fishing industry collapses, countless coastal towns and villages will become dead zones. The economic and social effects will be profound.

Just a few generations ago, this situation would have seemed unthinkable.

How have we allowed this to happen?

Our insatiable appetite for seafood, coupled with the brutal efficiency of our industrial fishing technologies, have wreaked havoc. But above all, it has been a combination of government weakness, industrial greed and a scientific community lacking the courage to sound the alarm that has resulted in one of the greatest ecological tragedies of our time.

Decision-makers have routinely ignored the warning signs. The reason is that they have been frightened of upsetting the 'fishing lobby'. As a result, they have set hopelessly unrealistic quotas, and have gone out of their way to appease industrial fishing companies. For example, in November 2008, the inappropriately named International Commission for Conservation of Atlantic Tuna (ICCAT) set a catch quota for bluefin tuna that is nearly 50 per cent higher than its own scientists advise.

Citing concern for jobs, livelihoods and consumer interest, politicians have brought fish stocks to the brink of collapse. And by their failure, they threaten the very people in whose interests they claim to be acting.

The truth, as any ordinary fisherman knows, is that this has never been a battle between fishermen and conservationists. Both want and need sustainable stocks of fish. It is a battle between two very different types of fishing: fishing communities and recreational anglers on the one side, and vast industrial fishing units on the other.

Consider the *Atlantic Dawn*. At 144.3 metres long, and weighing 14,055 tonnes, with a crew of 61, it is the world's biggest fishing vessel. It cost its Irish owner Kevin McHugh – who is a significant supporter of Ireland's governing Fianna Fail party – £50 million, and has purse seine nets, effectively giant marine shopping bags with drawstring necks, 3,600 feet in circumference and 550 feet deep. Its trawl nets are 1,200 feet in breadth and 96 feet in height. It can process up to 400 tonnes of fish a day and can store up to 7,000 tonnes of frozen fish, grossing about £1.2 million for each full fishing trip. So huge is the vessel that the Irish government had to encourage the EU to change its fishing rules to allow the *Atlantic Dawn* in European waters.

At the other end of the spectrum, over a million households in England and Wales have at least one sea angler. The total number is estimated at 3 million.

The fact that they all catch fish is the only common thread between them. Beyond that, they are necessarily in conflict. By choosing to identify the former, the factory fishing operations, as the so-called 'fishing lobby', our leaders have created a situation where fishing communities, countless livelihoods, the environment and food security itself are all under serious threat.

If instead the government chose to view the fishing communities that line our shores, as well as these 3 million anglers, as the true voice of fishing, the demands placed on them by this new and infinitely more authentic 'fishing lobby' would be considerably different, and considerably healthier.

Tools of Destruction

In a sustainable world, many of the tools of industrial fishing simply wouldn't exist. It is hard to imagine sustainable fishing while sixty-mile-long lines are permitted. Globally there are around 10 billion baited hooks on these lines, and each year they are responsible for the deaths of millions of sharks, hundreds of thousands of seabirds and marine mammals, and numerous endangered sea turtles.

The same could be said for purse seine nets. Some seines are a kilometre long and 200 metres deep and are enough to engulf two Millennium Domes if placed one on top of the other. They catch and kill entire shoals of fish – while trawl nets can kill up to 1,000 whales, dolphins and porpoises every day.

In the eighties, 'rockhopper' trawls were introduced. These are fitted with large rubber tyres that allow the net to pass easily over any rough surface, and in doing so destroy whole ecosystems – including coral reefs. The largest can move boulders weighing 25 tonnes. Scars up to 4 kilometres long have been found in the reefs of the North-east Atlantic Ocean. In heavily fished areas off southern Australia, 90 per cent of the surfaces where coral used to grow are now bare rock.

So-called 'ghost nets' – vast nets that have been lost or abandoned by fishing vessels – also drift through the oceans catching fish indiscriminately and causing havoc wherever they go. An estimated 1,000 kilometres of ghost nets are released each year into the North Pacific Ocean alone.

Where now?

Despite these horror stories, the good news is that it is possible to avoid catastrophe. Science and experience tell us there is still time. The oceans have a remarkable ability to recover when the pressure is eased. And quickly. When World War Two prevented fishing in the Atlantic, fish populations soared. The technology and the ideas are all there. What is lacking – as so often – is the political will. But given that a solution to this problem would be welcomed by the vast majority of fishermen and conservationists alike, there is no huge conflict to resolve, and no political courage is required.

What's more, whereas the problem is a global one, we don't need to wait for unlikely international action. Most of the world's ecologically and commercially important marine environments are coastal, and therefore national.

In a genuine solution, there will always be winners and losers. But in this instance, no matter how the calculation is done, the winners will dramatically outnumber the losers. Besides, if politicians decide to yield only to the interests of the industrial fishing sector, there can ultimately be no winners at all.

For now, there is little sign of progress. Instead, politicians pin much of their hope on fish farms, hoping that as the seas are depleted, we will continue to find sustenance from artificial systems.

Fish farming – or aquaculture – is already a major source of food for the planet's population; it is also big business. Almost 60 million tonnes of edible seafood were produced by aquaculture in 2004 – 43 per cent of all seafood sold globally. China is the world's leading aquaculture-producing country, with almost 70 per cent of the world's total production and an average annual growth rate of 12.4 per cent. In Scotland, which has 95 per cent of the UK's aquaculture, salmon farming supports employment of some 10,000 people, 4,700 of whom live and work in the remote rural communities of the Highlands and Islands.

On paper, it seems a perfect solution. But fish farming is intimately connected with the future of ocean fish stocks. Farming carnivorous fish, for example, is dependent on the capture of wild-caught fish, many of which, such as the sand eel and blue whiting, have been in headlong decline. Salmon farmers, for example, need at least three pounds of wild-caught fish to produce one pound of salmon. Fish farms have also been implicated in the decline of wild salmon and sea trout stocks, particularly in the west of Scotland, due to diseases that are rife amongst fish penned in so close together. Meanwhile, the chemicals used in aquaculture routinely contaminate surrounding waters.

Fish farms are potentially part of the answer, but they require much higher standards, and the focus should be on farming fish that eat vegetable protein. The list of approved medicines that are used in aquaculture needs to be monitored continuously, and the wider impact of fish farming on landscape and community needs to be properly taken into account. It is a part of the answer, but in real terms it is far from the first item on the agenda.

The first step is the necessary reform of Europe's controversial Common Fisheries Policy. The CFP is widely seen to have failed fishermen and fish alike. It is estimated that over the past couple of decades, EU ministers have routinely set fish quotas 30 per cent higher than stocks can accommodate, against scientific advice. There are growing calls for it to be scrapped altogether, or for individual countries to withdraw. The CFP needs profound reform, but there is, unavoidably, still a need for some kind of overarching framework for fisheries in the EU and beyond.

The UK should seek to stay in the CFP, but attach radical conditions and be willing to withdraw if those conditions are rejected. It needs, for instance, to devolve power to the local level, and, more importantly, it needs to be governed by science, not politics.

Specifically, Britain needs to at least double its sovereign waters to twelve nautical miles. Technically, it did precisely this three decades ago. But the sovereign status is forever being renegotiated, and the government enjoys only partial autonomy beyond the first six nautical miles. Outside this area lie international waters that are allocated by the EU to the fishing communities of various nations. The immediate six miles, over which we have complete control, are not enough to sustain our traditional fishing fleets. As a result, the rules dramatically favour international factory fishing. If we asserted our rights over the full twelve miles, and other nations did the same, we would be able to establish new rules and initiatives in the direct interests of our own authentic fishing communities.

Protected areas

By far the most effective measure would be to divide the sovereign waters into a vast network of marine protected areas. There would be varying levels – with the most highly protected areas being entirely off-limits to human activity; the second level allowing recreational fishing and tourist activities which do not damage the environment; the third level allowing fishing with vessels of a limited size; and the fourth level allowing some extraction. Marine protected areas are by far the quickest and best way to recover fish stocks. Where they have been introduced, they have worked and have been embraced by local fishing communities.

In Kyoto, Japan, for instance, the proportion of large male snow crabs rose by 32 per cent after just four years of protection. In New Zealand's Leigh Marine Reserve, the most common predatory fish are six times more abundant in the reserve than outside, and in the Tawharanui Marine Reserve, there are 60 per cent more species in the reserve than outside. And while Spain has suffered particularly badly from overfishing, catches close to the Tabarca Marine Reserve were 50 to 85 per cent higher after six years of protection than elsewhere. In the Galician fishing village of Lira, Spanish fishermen are now campaigning for a local reserve of their own – the first time this has ever happened.

The evidence is hugely compelling, yet only a tiny proportion of the UK's seas – no more than 2 per cent – are protected against destructive overfishing. This is nothing like enough space for fish stocks to recover and breed, safe from the nets, lines and trawlers. According to Professor Callum Roberts, author of *The Unnatural History of the Sea*, and professor of marine conservation at York University, a minimum of 30 per cent of the sea needs to be completely off-limits to fishing if we're to reverse the decline. Only an area this size, he argues, will allow migratory fish species like the common skate to recover.

There are signs that world leaders are beginning to listen. In early 2009, President Bush announced the creation of the world's biggest marine reserve – an area of the Pacific equal in size to Spain. It was a final grand gesture, perhaps designed to counter his image as America's least environmental president. Either way, it's good news.

If we work with real people, the fishing communities and the sea anglers, the mythical problem of the terrifying 'fishing lobby' would disappear. These people need fish stocks to be healthy, and would directly benefit from conservation. Stocks would improve, and without competition from the monster trawlers, they would thrive.

If sustainability means drawing on the world's interest, not its capital, then the world of fishing is perhaps the clearest example. If the Canadian government had maintained balance in its handling of the great Newfoundland fishery, the country could have reliably drawn about 250,000 tonnes of cod each year, for eternity. But because it plundered the 'capital', stocks were ravaged and the capital itself has shrunk to just 1 per cent of what was once an annual yield. It's difficult to imagine a stronger case for a constant economy than in our management of the oceans.

In our overfishing of the seas, we have come up against a solid ecological law: limits. From the oceans, as from the planet as a whole, we can only take so much before we destroy what we depend upon.

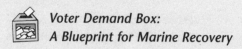

Voter Demand Box:
A Blueprint for Marine Recovery

Double our sovereign waters

We need to extend our sovereign waters – doubling them to at least twelve nautical miles. Doing so would allow the government and fishing communities to take a longer-term approach to managing fishing stocks, without fear of them being exhausted by others.

Marine protected areas

Working closely with the fishing communities, these sovereign waters need to be divided into different levels of marine protected areas. The government should lobby hard for the creation of a global network of MPAs, to recover stocks, protect the oceans and the fishing industry as a whole.

Ban the tools of destruction

These tools of destruction – the sixty-mile-long lines, the trawl nets and 'rockhoppers', the purse seine nets and the practice of discarding obsolete nets – are fundamentally incompatible with sustainable fishing, and indeed a sustainable world. They must be banned.

No more waste: limit the catch

Industrial fishing generates about 25 million tonnes of unwanted 'bycatch' every year. These unwanted fish are simply dumped overboard, dead or dying. For every pound of wild-caught shrimp, at least ten pounds of other sea life is wasted in this way. One clear mechanism for stopping this is to put limits on the total catch, irrespective of what it is that's been caught. That would encourage much more selective and careful fishing practice.

A new fishing lobby
Britain's sea anglers should be helped to form a large body of 'self-regulators' who would effectively police the newly established marine protected areas.

Higher standards for fish farms
The development of plant-based substitute feeds should be a priority. The list of approved medicines that are used in aquaculture needs to be monitored continuously, and the wider impact of fish farming on landscape and community needs to be properly taken into account.

Step Six

An Energy Revolution

We have a temporary aberration called 'industrial capitalism'
which is inadvertently liquidating its two most important
sources of capital: the natural world and properly functioning
societies. No sensible capitalist would do that.

Amory Lovins

Current energy policy defies all logic. In fact to call it a policy
at all credits it with a cohesiveness that it sorely lacks. Energy
– the way it is generated and the way that it is used – has been
a contentious issue for decades, and as a consequence, much
of what makes up Britain's energy strategy is a ragbag collec-
tion of ideas that do not address the fundamental challenges
that we face.

And, as a country, we are well aware of what lies ahead. The
government's former chief scientific advisor Sir David King has
described climate change as 'the most severe problem that we
are facing today, more serious even than the threat of terrorism.'
The Archbishop of Canterbury has called climate change a
'moral challenge' for humanity. And the Pentagon, in a secret
report leaked to the media during the presidency of George
Bush, envisaged a climate crisis in the next few decades,
concluding that climate change 'should be elevated beyond a

scientific debate to a US national security concern'.

But climate change is by no means the only reason for a rethink of our energy policy. Security of our energy supplies is a more immediate concern.

According to PriceWaterhouseCoopers, Europe needs to spend £5.6 trillion on improving its 'crumbling' energy infrastructure just to ensure a continuous future supply. When nine oil and coal-fired power plants, along with four ageing nuclear power plants, close in 2015, Britain's business sector will be crippled by an energy gap. The chief executive of National Grid, the company that operates Britain's power and gas transmission network, has warned that when this happens, unless the government acts now, Britain will face regular blackouts.

Ensuring that we have enough power to function as a nation is an absolute necessity, but doing so is not as easy as it once was. In 2005, Britain became a net natural oil and gas importer for the first time in over a decade. From 2020 onwards, on current projections, we are likely to be importing up to 80 per cent of our energy. The effect is that we are becoming unnecessarily vulnerable to international events.

Experts might argue whether oil has already reached its peak production, but everyone agrees that if it hasn't, it will. When it does, our dependence on oil – and on the countries producing it – will be severely tested.

If oil were to peak suddenly and supplies tail off more rapidly than expected, this would wipe out most if not all plans on offer from corporations, finance ministries and related organizations across the world. The very fabric of our society depends on readily available oil. Whether for transport, food, clothing or consumer goods, we need a constant supply. It is no exaggeration to say that our entire economic system and way of life is based on the assumption of ever-available cheap oil. If supplies are compromised – whether as a result of geological or geopolitical problems – it's hard to imagine the upheaval that would follow.

Faced with unexpected drops in domestic oil, there are strong indications from recent history that oil- and gas-producing nations would retain most or all of their domestic resources for their own use. For Britain, such a development would turn a national energy crisis into a catastrophe for which there is no precedent, one that would threaten our way of life forever.

This is not the stuff of conspiracy theorists. Warnings by oil industry insiders have recently reached a new pitch that should be sounding alarm bells in every capital in the world. When the CEO of Total, the ex-head of exploration and production of Saudi Aramco – the world's largest oil company – and the International Energy Agency separately suggest that we are at, or are approaching, the peak, we cannot afford to ignore them. As former US energy secretary James Schlesinger said, 'We can't continue to make supply meet demand much longer. It's no longer the case that we have a few voices crying in the wilderness. The battle is over. The peakists have won.'

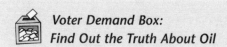

Voter Demand Box:
Find Out the Truth About Oil

A cross-party taskforce should be established immediately to draw up a risk assessment. It should not invite the traditional fuel industry to take part, as it would effectively be studying a risk scenario that says their maths is incorrect. The taskforce should be required to publically report its findings within a year.

At the same time, we should also expect our government to put pressure on the UN or International Energy Authority to undertake a review of the world's oil reserves. If the economic models of every nation on earth are based on the assumption of everlasting oil supplies, it is reasonable that they should know how much oil actually exists.

A new model

As long ago as 1990, Margaret Thatcher pointed out that

> many of the precautionary actions that we need to take would be sensible in any event. It is sensible to improve energy efficiency and use energy prudently; it's sensible to develop alternative and sustainable energy sources; it's sensible to replant the forests which we consume; it's sensible to re-examine industrial processes; it's sensible to tackle the problem of waste. I understand that the latest vogue is to call them 'no regrets' policies. Certainly we should have none in putting them into effect.

Sensible though these policies may be, it does require a government and an electorate to view change positively, rather than running scared from what seems alien to the current system. It's true that if our response to the energy crisis is the right one, the energy system of the future will look quite different to what we're used to. Yes, the effects will be profound, but getting there is less of a challenge than we are led to believe. And innovation in low-carbon technologies will pay off in any case, as carbon is priced into the market.

Rethinking how we generate energy is of course crucial, but

it's equally important to ask how we use that energy once it is generated. As a nation we are incredibly wasteful. In the home especially – an issue discussed in Step Eight – but also within the workplace. For example, there are 11 million industrial motor systems in the UK, which account for a staggering 40 per cent of the country's electricity consumption. They can account for two thirds of all power used on an industrial site. Simply including variable speed drives – which match electricity demand to the speed of the motor – can eliminate waste energy by up to 60 per cent. According to the Carbon Trust, British businesses could collectively save up to £2.5 billion in the next twelve months by following such cost-effective efficiency measures.

Save energy, save money

Six companies – IBM, DuPont, BT, Alcan, NorskeCanada and Bayer – have each reduced emissions by at least 60 per cent since the early nineties, collectively saving more than £2.4 billion in the process.

DuPont began a carbon dioxide and energy reduction programme ten years ago that today has brought greenhouse gas emissions down 70 per cent; in the same period, production increased almost 30 per cent.

The company claims it has saved £1.2 billion in efficiencies. BP, meanwhile, has reported savings of £394 million from emissions reduction efforts, and IBM reports savings of £479 million.

Waste is not always the fault of business and individuals. The way that we distribute energy is staggeringly wasteful too. Most of our electricity is produced using fossil fuels, and the way it is transported to the end-user is antiquated. Huge, centralized power stations transmit energy along transmission lines to houses and offices, yet only two thirds of the energy originally generated reaches its intended recipient. The rest is lost through the lines and as waste heat from cooling towers. A fifth of the UK's emissions of carbon dioxide come from this waste heat alone.

Before we begin to ask questions about how we will generate our energy, we should ask how we can use it more efficiently. If we can cut down on wastage and inefficiency we will need to generate less energy in the first place. The best power station of all is one that is not built, because there is no need for it.

Most current generation capacity is located too far from the consumer for waste heat to be used; switching to a decentralized energy system could radically change this – particularly if combined heat and power (CHP) plants were implemented.

The Danish Model

Developed in response to the seventies' oil crisis, the Copenhagen district heating system now accounts for over 50 per cent of Copenhagen's space heating, sending hot water through 1,300 km of insulated pipes around the city. The system's flexibility enables efficient use of fossil fuels while increasing the use of renewable energy and making communities more resilient to fuel price fluctuations. It connects four CHP plants, four waste incinerators and fifty small peak load boilers. The system costs 50 per cent that of conventional individual oil-based boiler and saves householders around £1,100 a year compared to conventional energy.

The federal government introduced tax incentives on fuel for such plants in the eighties, which means that people pay less fuel tax if they use CHP. Amendments to planning regulations have enabled municipalities to dedicate certain areas to district heating, and made it mandatory for households to connect to district heating. This led to an almost hundred per cent take-up, reducing both costs and emissions.

Today about 61 per cent of all of Denmark's households are heated by district CHP heating.

CHP is applicable on a variety of scales, from city-wide development down to individual buildings. Steady heat and power loads will improve the economics of CHP and so systems should be designed to allow a suitably sized engine to run at or near maximum capacity for as much of the day as possible.

CHP plants such as the ones in Copenhagen can be as much as 90 per cent efficient compared to the old-fashioned power plants we presently use in the UK, which are, on average, 38 per cent efficient.

In a CHP system, energy can be produced in the same way as conventional electricity, but the heat is retained for heating, hot water and cooling, and is distributed to customers via highly insulated pipes. This improves the overall efficiency of energy conversion to around 85 per cent.

A conventional CHP system uses natural gas to drive an internal combustion engine. It reduces carbon dioxide emissions compared to conventional distributed gas or electricity by 20–40 per cent. Some of the heat can also be used to provide cooling via absorption chillers.

If a small-scale, clean, efficient power plant is constructed near where the energy is to be used, the heat that would otherwise be wasted can be used. Given that in 2000, 83 per cent of energy usage in the home was for space and water heating, according to the Energy Saving Trust, the potential to use CHP is vast.

The World Survey of Decentralized Energy showed that 24 per cent of electricity output from newly installed power generation plants in 2005 came from decentralized energy systems, while the Netherlands, in little more than a decade, has made combined heat and power the single largest supplier of the country's energy needs. CHP also improves the resilience of the power supply. A general power cut or interference could only, for example, affect a very small area – something that New York discovered when the East Coast grid failed in 2003. Whole cities were plunged into darkness, all except the New York skyscrapers that had their own decentralized energy systems.

A Start in Britain

Woking is at the forefront of decentralized energy in the UK. Woking Borough Council has pioneered a network of over sixty local small-scale generators, including cogeneration (using waste heat to produce energy) and tri-generation plants (provides cooling as well as heating from waste heat), photovoltaic arrays and a hydrogen fuel cell station, to power, heat and cool municipal buildings and social housing as well as town centre businesses.

By doing so, the council has cut energy use by nearly half, and carbon dioxide emissions by 77 per cent since 1990. In the long term, doing so was no more expensive than the previous system. Generators are connected to users via private electricity wires owned and operated by Thameswey Energy Ltd – which is wholly owned by Woking Borough Council.

Woking raised capital for the development through energy-efficiency savings, and a fund mechanism was established for energy expenditure. Against this, savings from energy-efficiency measures were recycled, year on year, into further energy-saving initiatives. The large financial savings allowed the council to invest millions in energy supply innovation. Thameswey Energy has now also attracted investment from Danish pension companies interested in the low-risk return.

By developing a private network, Thameswey Energy also avoided normal charges from the use of the grid. These cost savings went to fund wires and generation to deliver low-emission electricity in competition with conventional suppliers. Woking shows that renewable technologies and cogeneration are highly complementary. The key lesson is that, liberated from the constraints of centralized rules and infrastructure, decentralized energy can be extremely competitive and lead to large savings in emissions.

We need an energy revolution. We can do it, as others have demonstrated, and when we do, we will see a huge range of new opportunities.

 Voter Demand Box:
Capture the Heat

When power plants are located too far from consumers, the heat they generate is wasted. The best way to remove this waste from the system is to introduce a levy on waste heat from future electricity generators. Such a levy would immediately encourage companies to base their power generating capacity as near as possible to where it is needed, and facilitate the move towards community-level generation. Over the next few years, many of our ageing power stations need to be replaced. Having such a levy in place when this happens would help transform the prospects for renewable, decentralized energy in the UK. The money raised via the levy would need to be used to invest in a decentralized energy infrastructure.

The home energy pioneers

Over the last few years, UK businesses have successfully used government support to develop onshore wind farms – one of our few renewable energy success stories. But we have been slow to develop other forms of renewable energy such as solar, marine and decentralized energy. Other countries are blazing a trail. Ground source heat pumps, for example, which transfer heat from the ground into buildings, are by now the most common heating system in single-family houses in Sweden.

The Energy Saving Trust believes micro-generation could provide 30 to 40 per cent of the UK's electricity by 2050 and reduce domestic emissions by around 15 per cent. Given the massive private investment flowing into clean energy, this seems likely. According to New Carbon Finance, global investment in new energy alone amounted to £61 billion in 2006 – up from £42 billion in 2005. Ernst & Young forecasts growth to £455 billion in the next decade.

But this still remains much less than the money flowing into conventional energy. One way to tip the balance is to follow the German example of rewarding homeowners for generating their own energy. It's called the Feed-in Tariff (FIT), and despite the name, is a simple concept. Anyone generating solar photovoltaic, wind or hydroelectricity is guaranteed a twenty-year fixed payment at a level designed to cut payback time to a matter of years. Electricity generated by solar panels, for instance, gets a tariff six times greater than that paid for electricity generated by large-scale wind. The system boosts take-up by consumers by reducing the payback times on such investments to fewer than ten years – compared with twenty-five or thirty years in Britain – and gives industry the certainty of long-term demand to make it worthwhile investing in new technologies and generating plants.

The results have been spectacular. Germany has 200 times as much solar energy as Britain. It generates 12 per cent of its electricity from renewables, compared with 4.6 per cent in Britain. The industry has also created a quarter of a million jobs – a number that is growing fast. In stark contrast, Britain has only 25,000, a number that represents the amount of jobs created in the industry in Germany in 2008 alone. The Germans introduced the FIT in 1999 and tweaked it in 2004. Since then, the idea has taken off globally. FITs have now been adopted in nineteen EU countries, and forty-seven worldwide – but not in Britain. German renewables firms are now world-beaters and the German economy has been strengthened, not weakened, by a rush into renewables.

In Britain we use a different system, and it is widely seen to have failed. It's called the Renewable Obligation, and requires energy providers to buy a percentage of their energy from renewable sources. Because onshore wind is currently the least expensive, virtually all the money flows to what is a relatively mature technology, and much less in need of subsidies than many other forms of new renewables.

The beauty of the German system is that it is flexible. It doesn't depend on governments to pick technology winners, something most governments are very bad at doing. It simply helps new technologies onto the ladder and rewards home-owners for using them.

Freiburg, a town of 200,000 people in the Black Forest, has almost as much solar photovoltaic power as the whole of Britain. By the time Britain starts constructing its first 'eco-town' in 2016, Germany will have fifty or sixty eco-cities. Small wonder that the Labour government has quietly dropped the pledge it made six years ago to catch up with Germany by 2010.

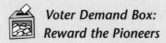

Voter Demand Box:
Reward the Pioneers

For encouraging the growth of decentralized energy systems, in particular, the Feed-in Tariff system has been hugely effective. A FIT fixes the export price for electricity at a level necessary to stimulate investment without the price uncertainty caused by competitive bidding. This policy has also had a dramatic effect in other European countries. We should demand that our government bring it to Britain.

Planning for the revolution

Currently, some forms of renewable micro-generation, such as CHP boilers and biomass generators, can be installed within homes and businesses and not be subject to planning constraints. The same, however, cannot be said for other forms of micro-generation, such as solar panels and mini wind turbines. These can remain subject to planning rules.

Giving local authorities the right to set their own eco-standards for buildings is crucial. The so-called 'Merton Rule', introduced in 2003, requires any new buildings to reduce their carbon emissions by 10 per cent through the use of renewable energy sources. On the back of the rule, Merton Council won several national awards.

117

> ### Voter Demand Box:
> ### Break the Rules
>
> The installation of micro-generators should be classified as a 'permitted development' – essentially a fast track to planning permission – and local councils should be given the right to go beyond existing norms in encouraging micro-generation. Planners should be able to require applicants for new development schemes to include plans for local generation of energy.

A clean energy infrastructure

When we talk about renewable energy in the UK, we need to be realistic. Solar power isn't going to generate enough power for our needs. But thanks to the 7,760 miles of coastline around the UK (excluding the islands), we have the best offshore wind, wave and tidal assets in Europe. Yet we once again lag behind most other countries in Europe. When we consider that 1.2 million homes are powered by wind, ten times less than in Germany and five times less than in Spain, it seems a shocking waste of resources. The government itself believes that at least 25 per cent of our electricity could come from wind by 2025, and a recent report is even more bullish, estimating that wind could provide around 45 per cent of our annual electricity. Of this, 90 per cent would be from offshore turbines. Add to that the power in our seas – which has the potential to provide 12.5 per cent of our electricity by 2025 – and the case for clean energy projects becomes increasingly compelling.

One innovative example is the Wave Hub, a ground-breaking £28 million wave power project ten miles off the coast of Cornwall that uses seabed power cables to connect a network of wave power generators just below the surface to land. The hub will generate electricity for 7,500 homes and bring 1,800 jobs and £560 million to the UK economy over twenty-five years. Almost 1,000 of these jobs and £332 million will be generated in south-west England.

Germany, meanwhile, is putting serious thought into the possibility of a macro-grid which would link together the huge solar energy potential of the Sahara, Iceland's geothermal energy, hydropower in Scandinavia and the Alps, and wind power in the North Sea. Such a scheme raises huge difficulties, but is without a doubt worth exploring.

There is no escaping the fact that the use of coal will continue to rise. For one thing, there is plenty of it. For another, China has no plans to stop using coal. In 2008 it added thirty new coal power plants to the 550 it already has, and if trends continue, it will have 800 by 2020. India, meanwhile, built forty in 2008 alone.

Coal is a very carbon-heavy and polluting form of energy releasing 29 per cent more carbon dioxide than oil and 80 per cent more than gas, on average. Drax coal power station in Yorkshire's 'Megawatt Valley' may be the UK's cleanest coal station but it still annually puffs out the same volume of carbon dioxide as just over a quarter of Britain's cars, or nearly a third of our homes.

Whilst new 'clean coal' technologies are emerging, even the cleaner coal stations still have efficiencies of only around 40 per cent, a paltry amount when compared to the 90 per cent of some biomass and CHP plants in Denmark. The main technological development that could ensure a future for coal in a genuinely low-carbon world is carbon capture and storage (CCS). CCS captures carbon from combustion and pipes it into geological formations or old oil and gas beds under the sea. It's a solution that its backers believe will cut emissions by up to 85 per cent.

To make this happen, and at scale, the coal sector needs clear signals from government. In the UK, we're miles away from that. In contrast, California Governor Arnold Schwarzenegger has made it explicitly clear that all new coal plants in his state must use CCS. He has ensured this will happen by insisting that all new power plants serving his state must meet new efficiency standards. The key requirement is that no plant can be more polluting than a modern gas-fired power plant, by far the cleanest of the hydrocarbon technologies. He hasn't instructed the energy firms how to achieve this – that's up to them – but just as makers of household appliances have been set energy-efficiency standards, and have delivered, so too will the energy firms.

One of the main reasons that the UK lags behind other countries in its use of clean energy technologies is the shameful record of investment – or lack of it – by all governments over the past two decades. If the government is serious about using clean energy to tackle climate change – which it insists that it is – then it must put its money where its mouth is. And as a first, and crucial step, that means investing in an improved energy grid.

Whereas the private sector is best placed to deliver energy from a variety of sources, it is for the government to provide the grid. For example, on our coasts, without investment in underwater cables, neither wave power nor offshore wind will flourish. Without a long network of pipes to carry the captured carbon, CCS won't be possible. Just as all power plants benefit from the national grid, without which they couldn't distribute their product, so too will CCS plants require the infrastructure to be provided for them.

Biogas too could massively reduce our dependence on gas

imports, but only if the infrastructure for collecting and distributing it is in place. Biogas is created by decomposing organic waste, and contributes less than 1.5 per cent of current gas use. But according to the National Grid Company, with the right infrastructure, biogas could provide us with half our gas needs. Other studies suggest that the EU could use biogas to easily replace all of its imports from Russia by 2020.

Perhaps the most significant part of President Obama's green energy plan is his proposal to invest in a 'smart grid'. Using modern digital technology, it would coordinate supply and demand, and allow two-way communication between energy suppliers and consumers, enabling utility companies to direct power more efficiently away from low-energy users to high-energy users depending on the time of day or need. It would also allow suppliers to sell cheaper energy during times of low demand, at night for instance. Consumers could save money by charging their cars or setting their washing machines when energy is less expensive. It's an appealing idea, but the cost of the upgrade is more than the budget Obama has put aside for it.

 Voter Demand Box:
Invest!

We urgently need a renewable energy fund to provide substantial grants for the research and development of radical new clean energy technologies. From wave power to clean coal technology, potential solutions remain in the pipeline due to a lack of investment. Government should provide that investment. Diverting money that would otherwise be spent subsidizing fossil fuels or the nuclear industry could provide billions of pounds for research, support and, crucially, for upgrading the national grid.

Paying the polluters

Wind farms, solar power plants and any number of other renewable energy projects cost money. It's a rod that opponents of renewable energy use to beat down any hope of progress. But if it is costs to the taxpayer that we're looking at, then the subsidies provided by governments all over the world to the fossil fuel industry must be taken into account. Huge sums of public money are spent each year to prop up a form of energy generation that is wrecking the planet.

The estimated global public subsidy for fossil fuels is between £91 billion and £152 billion a year. Subsidies for oil products in non-OECD countries are estimated at over £55 billion annually. The European Environment Agency has produced an analysis of energy subsidies which concludes that for every £1 spent on subsidizing renewable energy projects, £4 is spent on fossil fuels. Another report states that every year the British government spends some £6–8 in fossil fuel subsidies for every £1 it spends to support clean and renewable energy.

The former head of BP, Lord Browne, recently called for the dismantling and reallocation of these subsidies, saying that renewables cannot compete unless there is a level playing field.

But what do these 'subsidies' involve? For a start, the oil and gas industry in the UK is supported by keeping VAT on its product artificially low – at a cost to the Treasury of around £685 million a year. A significant amount of the UK's support for fossil fuel projects is in the form of export credit guarantees, through which public money is used to underwrite investments in large-scale energy projects in the developing world. Diverting a small fraction of this money towards renewable energy projects or energy-efficiency incentives would make an enormous difference.

Very little of what tax the government does get from oil and gas is invested in developing alternatives and energy efficiency. The revenue that government collects from the fossil fuel sector may be greater than revenue from council tax, stamp duty, capital gains tax and inheritance tax combined. Compare that to Norway, which set up a substantial fund to invest the proceeds of oil taxation to ensure that future generations would benefit once oil has gone. At the end of 2005 that fund stood at £127 billion – the equivalent of £27,000 for every Norwegian. The UK should establish a similar fund.

Elsewhere, attempts to end government support for destructive fossil fuels are already underway. The 'End Oil Aid' Bill, introduced in the US in April 2007, seeks to end government support for the international operations of oil companies, calling on international financial institutions to stop financing oil and gas projects. Calculations by the World Bank and the OECD show that removing such support globally could reduce carbon dioxide emissions by around 10 per cent worldwide. Similar proposals were put forward in 2001 at a meeting of the G8 in Genoa, where a report commissioned by member countries called on nations to 'remove incentives and other supports for environmentally harmful energy technologies'. It also encouraged them to shift the priorities of international lending agencies like the World Bank to support more clean energy projects in poor countries. The motion was rejected by the US.

> ### Voter Demand Box:
> ### Stop Paying the Polluters
>
> **Our government should draw up a roadmap for phasing out subsidies that are harmful to the environment, and should encourage the European Commission to do the same.**
>
> **Any lending available from export credit agencies and multilateral development banks should be directed towards investment in renewable energy and energy-efficiency projects only. The government should apply pressure on the World Bank to do the same.**

Why not nuclear?

Before climate change forced its way onto the mainstream political agenda, nuclear energy was in terminal decline – hounded by environmentalists, rejected by investors and mistrusted by the general public. Then, all of a sudden, it was presented as a solution to the greatest threat we've ever faced – global warming.

While the benefits of nuclear power have been reassessed, its dangers have not suddenly gone away. In fact, if anything, they have multiplied. In the post-9/11 world, no one would claim, as the director of the French nuclear installation giant, COGEMA, did shortly before the events of that day, that a nuclear power station was not at risk from an airborne attack as planes are 'forbidden to fly over it at low altitude'. As Director General Mohamed ElBaradei of the International Atomic Energy Agency warned, 'If the terrorist is willing to die, that changes the security equation drastically.'

When two Greenpeace activists dressed as missiles and

carrying suitcase 'bombs' managed to access Sizewell B in southern England, they were proving more than a point. They were demonstrating that nuclear power stations are obvious targets for radicals. Their fears were borne out when detailed plans of the same plant were reportedly found in a car connected to the bomb attacks on London's transport system in 2005. In 2007, meanwhile, Australian police stopped suspected terrorists who were believed to be staking out a nuclear research reactor near Sydney. It is worth considering that an attack on Sellafield in Cumbria would be forty times more disastrous than the Chernobyl accident, according to the EU Commission.

That said, while the security dangers are large, they pale next to even the most conservative climate change predictions. If nuclear power genuinely offered a solution to climate change, the argument against its use would be null and void. But the truth, as the Sustainable Development Commission discovered following its own investigations, is that the contribution of nuclear to emissions reductions has been wildly exaggerated. Even if we replaced our existing nuclear reactors — and doubled their number — we'd see an 8 per cent reduction in carbon emissions, and not until 2035. Neither can nuclear power stave off the looming energy crisis we face in this country.

Nuclear energy provides just 19 per cent of our electricity, or just 3.5 per cent of our total energy needs in the UK. Even its supporters are only talking about replacing existing nuclear capacity, so its contribution will remain small, and in the discussion about energy security, it is a red herring. Nor is nuclear power immediately available. We have something in the region of ten years, at most, to replace existing capacity. But if history is the indicator, nuclear takes longer. The planning process for Sizewell B started in 1981, and it wasn't until 1995 that the first power was produced. That's not the case with energy efficiency, which can happen immediately, or renewable technologies like wind, tidal or solar, which all have shorter timescales for implementation than nuclear.

If we put these issues aside and assume it is possible to build new nuclear capacity in record time, there is still the issue of cost. Nuclear is hugely expensive. As *The Economist* magazine has remarked, 'nuclear power, which early advocates thought would be "too cheap to meter", is more likely to be remembered as too costly to matter'. It's not just the waste that costs, although that's a significant problem. In Britain alone we will need to find somewhere between £56 billion and £70 billion just to deal with existing nuclear waste.

But it's more than that. The whole operation requires continued state support. It is a fact that there has never been a fully privately financed nuclear reactor in the history of the industry. Even Sir David King, formerly the government's chief scientific advisor and a keen advocate of nuclear power, has said that 'for the nuclear new-build industry to go ahead, they would need some assurance that they would not be faced with short-term competition'. In the US, where no new nuclear reactors have been built for thirty years, the government has announced a generous new subsidy programme. Provisions include: production tax credits for up to £76 million per 1,000 MW, federal loan guarantees covering up to 80 per cent of project costs, and up to £300 million in risk insurance per plant in the case of delays in the plant becoming operational.

A decision in favour of nuclear energy diverts money away from genuinely cost effective solutions. Take energy efficiency. According to the Rocky Mountain Institute, a pound invested in energy efficiency buys seven times more 'solution' than a pound invested in nuclear. To put it in context, if every light bulb in the UK was exchanged with an energy-efficient model, we'd save the power equivalent of nearly two advanced gas reactors. We also know that retrofitting old homes could lead to a 60 per cent reduction in carbon dioxide from the housing sector by 2050. Much of this has already been proven by companies like Bayer, BT, DuPont and NorskeCanada, that have reduced their greenhouse gas emissions by at least 60 per cent since 1990 with total gross savings of £2.4 billion. In terms of value for money, there is no contest.

The existing nuclear infrastructure is crumbling and needs replacement, so the time for decisions is now. But whatever course we choose we will have to stick to it for years to come. The opportunity for positive change is immense. We must take it.

Rapid change is possible

When faced with gloom and pessimism, clean technology guru Professor Amory Lovins often cites this story. By the early 1850s, he explains, most homes in the US were lit by lamps that burned whale oil. Before long, demand began to exceed supply, and the whales became harder and harder to find. Prices rose steeply, and alternative sources of energy like smokeless coal-kerosene began to take hold. By 1859, when Edwin Drake struck oil in Pennsylvania, five sixths of all whale-oil lamps had switched to the new fuels. The astonished whalers, who hadn't heeded the competition, ran out of customers before they ran out of whales. Rapid change is possible. Indeed it's inevitable, but we must make it happen on our terms.

Step Seven

Getting Around

And that will be England gone,
The shadows, the meadows, the lanes,
The guildhalls, the carved choirs.
There'll be books; it will linger on
In galleries; but all that remains
For us will be concrete and tyres.

Philip Larkin, 1972

On a perfect summer's day in Paris 1895, the first motorized road race began. The streets were swelled with onlookers as twenty-seven entrants were waved off on their journey to Bordeaux and back. It was not only a battle between drivers, but between ideals. The majority of cars were propelled using petroleum engines, but many steam enthusiasts were convinced that their method of propulsion would prove the better. Whatever the outcome, one thing was certain: our love affair with the car had begun. When the eventual winner – 53-year-old Emile Levassor – crossed the finish line at the Porte Maillot, something radical happened. As Ruth Brandon says in her book *Automobile: How the Car Changed Life*, 'the car took its place as the ultimate consumer good'.

Over the course of a century, the car has gone from a rich man's plaything to a commodity that the vast portion of the population needs to function on a daily basis. As cars became more popular, and more affordable, society quickly began to shape itself around their existence. Whole cities, towns and areas are predicated on the assumption that residents have access to a car. What was once a great luxury has become a necessity, and more recently, a requirement.

For many, driving remains a passion and an enthusiasm. *Top Gear* has become one of the BBC's flagship shows. But beyond showcasing the cars themselves, what the programme – and its ever-widening net of merchandise and competing programmes – actually provides is a fantasy idyll for the passionate driver. A fantasy land that will allow them to push a Ferrari over the 150 mph mark with impunity, that provides thrilling chicanes to test even the best racing suspension – in short a place that banishes the appalling reality of driving on Britain's clogged and congested roads

Our roads are so packed that it has become impossible to build new ones fast enough to take the strain. There are 34 million registered vehicles on our roads today. Roughly 28 million of them are cars, and that number is growing. If we all tried to access our towns by car they would be unable to cope with the gridlock. Traffic in London today is, notoriously, as slow as it was a hundred years ago, in the days of walking and horses. The car was supposed to make things so much better, so much more convenient for us, but it rarely turns out that way. The average Briton spends 235 hours in their car every year – much of it going nowhere.

Over the last half-century, our society has developed to be so car-dependent that in truth, they have begun to enslave rather than liberate us. The car has made possible all those out-of-town superstores, which in turn have led to all those new housing estates with no shops. It has allowed politicians to ignore bus services and cycle routes. And without them more people are forced on to the road, especially now that they can work ever further away from where they live. It's a vicious circle, which will take real fortitude to break.

Getting it wrong

For decades, the political response has been to frantically build new roads, or widen existing ones, in a doomed bid to build our way out of the problem. But it doesn't work. Not only do new roads destroy the countryside and local communities, it's practically impossible to build enough roads to keep up with predicted levels of traffic growth. Not only that but building new roads cannot help but attract new traffic. Traffic volumes on the controversial Newbury bypass, which opened in 1998 after huge protests, were already 22 per cent above the forecasted level for 2010 within six years. Peak-time congestion within the town is now back to levels that prompted the building of the bypass in the first place.

Clumsy transport policy causes more than just congestion. Overall, the transport sector contributes roughly a quarter of Britain's carbon dioxide emissions, the vast majority of that resulting from road vehicles. Road transport, meanwhile, accounts for about three quarters of Britain's fossil fuel use.

The train network in this country has been a national joke for decades. But the joke is increasingly wearing thin, as fares continue to rise well above the rate of inflation, while services are not improving commensurately. Compared to our European counterparts, the rail system in Britain is unreliable, confusing and simply too expensive to be an alternative to taking the car. This is especially true of trains that attract high numbers of travellers. Taking the train from London to Manchester – or vice versa – on Friday night can be a nightmarish experience of unreserved seats, packed carriages and stressed out members of staff trying to quell the growing unrest. It's no wonder, therefore, that people look to other methods of transport – especially air travel.

Just as our train infrastructure is burdened by too many passengers, so too are our airports, hence the government's vision for airport expansion. Their initiatives will treble the number of passengers moving through UK airports, rising to an estimated 470 million by 2030. Two million people already live beneath Heathrow's flight path and the quality of their lives is routinely undermined by the noise of low-flying aircraft. But its effects go well beyond West London.

Aviation is the fastest growing cause of greenhouse gas emissions. If the UK meets its target of reducing overall carbon emissions by 80 per cent by 2050, the government admits that emissions generated by Heathrow airport alone will account for about a fifth of the country's total carbon budget. Some studies even suggest that if aviation is allowed to grow unconstrained, it could account for all of the emissions we are allowed to generate in 2050 – requiring every other part of the economy to emit nothing at all.

Economic necessity is often used to justify airport expansion. More jobs, more opportunities, a wealthier community. But there are holes in the argument, not least the fact that the costs of aviation are routinely ignored. The £9 billion tax subsidy that accrues to the sector through tax-free fuel and zero-rated VAT isn't accounted for. Nor are the environmental and social costs included. The £15 billion tourism deficit – the difference between the amount spent by incoming tourists and British residents holidaying abroad – is also ignored. When the government claims that we will enjoy a net benefit of £5 billion from expansion of Heathrow, over seventy years, it is because it has ignored these costs. It fails even to account for the congestion that will be caused by an extra 25 million road passenger journeys to and from Heathrow each year, or the effect on house prices under the flight path.

These indirect subsidies, along with some creative accounting, flatter the economic case for airport expansion and stoke demand by reducing the cost of flying.

Still, the government fears that if we don't expand our airports, we will jeopardize the competitiveness of our aviation industry. But look at Heathrow. It is already the world's busiest international airport, serving 17 per cent more passengers than its closest rival, Charles de Gaulle, and 45 per cent more than Schiphol. And let's not forget that there are another four 'London' airports, which between them handle one third more passengers than Paris and nearly two thirds more than Amsterdam. Our competitors have a lot of catching up to do.

The truth is, we would do far better to focus on improving existing capacity. For instance, should we encourage so many transfer flights? One in three people using Heathrow never even leave the airport, and therefore add no value to the UK economy, a point made recently by the former chief executive of British Airways, Bob Ayling: 'Transfer traffic in its own right is loss-making. What the government doesn't see is that transfer passengers spend no money in Britain, at least little beyond the value of a cup of tea.'

It's not that opponents of airport expansion are killjoys, committed to preventing families their yearly holiday or seeing distant friends and relatives. No one is seriously suggesting the end of aircraft travel. Instead, they want to reduce unnecessary flights – short-haul flights, for instance, to places already reachable by train. Over a 500 km trip, aircraft emit six times more greenhouse gases than high-speed trains, and twelve times more than a coach. Around 100,000 of the 470,000 flights using Heathrow airport every year are to UK or near-

Europe destinations where there already exists a reasonable rail alternative. Paris is the most popular destination, with around sixty flights to and from Heathrow a day. There are more than thirty flights a day to and from Manchester. Yet fast and efficient trains already exist between these destinations, and with the right approach they could be dramatically improved.

We have a shorter distance of high-speed rail (HSR) track in the UK than they have in Belgium. Our meagre sixty-eight miles is barely a patch on France's 963 miles, or Germany's 545 miles. Eurostar trains now travel from London to Paris in just over two hours; half the time it takes by plane (if the entire journey is taken into account). Yet it currently takes around five hours to travel from London to Glasgow by train, and two and a half hours from London to Leeds. High-speed rail could reduce these journey times to two and three-quarter hours and one and a half hours respectively. The attraction of such speed has meant that wherever the HSR network has been expanded in Europe, that country's domestic air traffic has fallen.

A high speed rail network for the UK would cost around £36 billion – a vast sum, but one that when you consider its benefits is more than value for money. This is a system that would benefit all, and would encourage people off the road. Can the same really be said for the £5 billion price tag attached to widening 240 miles of the M1? As an investment, HSR offers Britain the opportunity to lead the way in public transport.

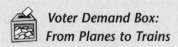

Voter Demand Box:
From Planes to Trains

Massive investment in high-speed rail
We need massive investment in the rail network to reduce journey times between major cities. Without a viable, reliable alternative, it is unreasonable to expect people not to fly short haul. But any reliable alternative will cost, and the money will not all come from the private sector.

Taxing short-haul flights where rail alternatives already exist is therefore an obvious means of raising the necessary funds. But we must insist that every penny of this new tax is ploughed back into the rail network.

Moratorium on airport expansion
The government plans to expand a number of UK airports. It is a nonsense to do this whilst talking of reducing our greenhouse gas emissions. The government needs to put a moratorium on the building of new airports, and expanding existing airports, and focus instead on improving existing capacity. This policy must be linked with offering alternatives to the traveller.

Where now?

Just as it would be futile to call for the end of aircraft travel, we need to be realistic about other methods of transport. There is no doubt that cars are here to stay. Over time, with the right kind of policies, their numbers could shrink, but until then we need to clean up the existing fleet.

American cars consume 8.2 million barrels per day of oil, a figure that nearly matches the 8.4 million barrels per day produced by Saudi Arabia – by far the world's largest oil producer. Cars and light trucks are responsible for 20 per cent of US carbon dioxide emissions. In the UK, it costs us £12.7 billion each year to run the 34 million vehicles on our roads. They emit an estimated 121.2 million tonnes of carbon dioxide every year.

We have the technology. If all existing cars performed as well as the cleanest vehicles in their individual class, their emissions could be reduced by up to a quarter with no other action necessary. If every UK vehicle was powered by clean electricity, we'd cut carbon dioxide emissions by more than a fifth, as well as reducing our dependence on other countries. The cost of running an electric car is anything between eight and sixteen times less than a conventional car, so for consumers, the benefits are obvious.

These are solutions that already exist, but with technology having the habit of rapid advancement, there's reason for greater optimism. It's important, though, that governments stay away from prescribing the solution: politicians generally get it wrong when it comes to picking technology 'winners'.

But what could be considered potential winners? We may see the emergence of hydrogen cars. Most of the big car makers are researching the possibilities and a variety of demonstration models already exist. But given the urgency of the challenge, and the likely time scale of hydrogen cars, electric cars are probably more realistic.

Electric cars are already a common sight in cities where journeys are short. But there's a risk over longer journeys that drivers will find themselves low on battery and miles from a power point. Companies like Project Better Place, a California-based start-up company, are developing answers.

Backed by the Israeli and Danish Governments, the company is finding ways to mainstream electric cars. Better Place essentially offers a 'battery service' much like a mobile phone contract. Consumers do not buy batteries. They buy into a contract that allows them to swap their depleted battery with freshly charged ones at any one of a large number of battery stations. The cost of driving is dramatically reduced, and all the reasons we wouldn't normally buy an electric powered vehicle, not least fear of being stranded, are removed. The government will support the initiatives through tax breaks on zero-emission vehicles. Similarly, in Denmark the government is planning to build an electric car network with about 20,000 recharging stations powered by wind.

The temptation for governments afraid of tackling car industry lobbyists is to promote biofuels – fuels derived from a wide variety of crops – as an alternative 'clean' source of fuel. On paper it appears to make sense, but in reality, biofuels have proven to be an ecological disaster and on every level a totally unrealistic alternative to oil.

For one thing, the maths doesn't come close to adding up. To produce enough fuel to fulfil our current global energy requirements we would need 10.8 million square miles, almost twice the world's available arable land. To run all of the UK's cars, buses and lorries on biodiesel would require 25.9 million hectares – almost five times Britain's total available arable land. According to Professor David Pimentel at Cornell University, if the entire US corn crop was converted into biofuels, it would replace just 7 per cent of petroleum consumption.

This in and of itself should be enough to show these kinds of biofuels as the white elephant they are. But there are other effects that make them even more impractical. The rush to biofuels has already had an impact on food prices. In a country where people spend just 5 per cent of their income on food, a 10 per cent rise in food prices isn't necessarily a problem. But where people spend up to 40 per cent on their food, such a hike is painful.

The environment is suffering too. Throughout the world, tropical forests are being destroyed to grow palm oil to produce biofuels. A 2005 Friends of the Earth report found that between 1985 and 2000, palm oil plantations were responsible for 87 per cent of Malaysian deforestation. 'We call it "deforestation diesel"', says Simone Lovera, managing coordinator of the Global Forest Coalition, an environmental NGO based in Asunción, Paraguay. 'Biofuels are rapidly becoming the main cause of deforestation in countries like Indonesia, Malaysia and Brazil.'

Not all biofuels are bad. At their best, they can make a contribution to the fight against climate change. The Canadian biofuels company Dynamotive, for instance, has been producing 22,000 tonnes of biofuel a year since 2005 from waste woodchips, construction waste and coffee-bean shells. They are now running a successful pilot scheme producing oil from human sewage to generate heat and power in diesel engines and boilers. These are second-generation biofuels: they use waste products to generate energy and as such, they neither displace habitat nor food production. They are the future of biofuels.

But they are only a small part of the overall future. Governments need to impose higher standards on the car makers, and use the tax system to cut the price of the cleanest cars. Given that people change their cars on average every six and a half years, the shift could be rapid.

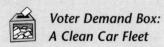

Voter Demand Box:
A Clean Car Fleet

Demand higher standards

A review commissioned by the UK Treasury has worked out that emissions from our roads could be halved by 2030 with a focus on cleaner, more efficient cars, using cleaner fuels. By 2050 reductions of up to 90 per cent are possible with battery electric systems using low-carbon electricity. The government – ideally working with other European neighbours – should set a far higher standard for cars. As long as standards are set realistically, car companies will deliver us the cars we demand.

Make green cars cheaper

Instead of introducing penalties for those of us who have already made our choices, the government should be encouraging change where it matters – at the point of purchase. The disincentives need to be matched by the incentives. To trigger a rapid shift in the quality of our collective car fleet, the government should introduce a high 'purchase tax' on the biggest, dirtiest cars and match it pound for pound with tax relief on the cleanest cars. If the incentive to purchase a clean car is big enough, the UK could have a clean car fleet within a matter of years. Similar systems already work elsewhere, for instance in Austria, Denmark, Sweden and Holland.

Government buying power

The British public sector operates the largest fleet in the UK, with well over 300,000 vehicles. Government departments should buy only the most efficient models.

Given that the British government spends about £2.2 billion every year on the public car fleet, there are huge savings to be made, both financially and in terms of emissions.

Clear standard for biofuels

We urgently need a stringent international standard for bio-fuel production. Such a standard should discriminate against biofuels which generate no net carbon saving, are generated at the expense of valuable habitat, or are generated at the expense of food production and security.

Ending the car culture

Cleaning the car fleet can happen quickly, but it does nothing to address the deeper-seated problems facing transport in this country. The engrained car culture – a culture of dependence – needs to be challenged. This means abandoning the inadequate 'predict and provide' approach to new major roads and replacing it with a sustainable, integrated transport strategy. In other words, through better planning and design we need to reduce the need to travel in the first place, and to make it easier to complete the journeys that remain using alternatives to the car.

As we build new settlements, and improve existing towns and cities, we should do so with a view to designing out car dependency. People need to have the option of visiting shops, leisure facilities and workplaces close to their homes, which means building mixed-use developments instead of so-called 'dormitory towns' where people reside at night, but rarely during the day. Where money is invested in transport, there should be a bias in favour of alternatives to the car, for instance rail, buses, walking and cycling. People should have the choice *not* to drive, rather than have a car journey forced upon them by bad design. The effect would not only be less pollution and congestion, but also stronger and more vibrant communities.

Britain's public transport needs to step up to the plate. We

139

need to be both bold and brave in our attempts to lower carbon dioxide emissions and reduce congestion. This is not just for the good of the environment, but for the good of our health. The morning journey to work is surely the most stressful part of a commuter's day, and any way that we can alleviate this will be hugely beneficial to the country.

On the roads, we need, among other things, better buses. In London, car ownership has declined as a direct result of the promotion and funding of good bus services. In York, a package of measures designed to alleviate congestion in the city centre has meant that journeys are now faster by bus than by car. In Leeds, Brighton and Hove, town centre car traffic has decreased 10 per cent over three years and bus use is growing steadily. In Cambridge, bus use has grown by a remarkable 70 per cent. These schemes prove that when given the option, buses can be popular. But if they are to really change the car culture of Britain, they need to be deployed where they can make the most amount of difference: the school run.

In the US, Yellow School buses represent the largest mass transit system in the country. Some 450,000 take more than 25 million children to and from school. Each school bus takes between thirty and sixty cars off the road during peak rush hour. The leading US school bus manufacturer, IC Bus, is producing the nation's only line of hybrid school buses, which improve fuel efficiency and reduce diesel emissions by up to 70 per cent. The school buses are 'plugged in' at night, charging the electric motor that is part of the hybrid system. Each hybrid school bus is estimated to save £1,800 annually, and 800 gallons of fuel. Our roads and our environment – not to mention commuters – are crying out for such a scheme to be introduced across the UK.

And efforts to cut unnecessary journeys do not end with other motorized solutions. Most car journeys in the UK are less than five miles. In Lyons, for instance, the city's so-called 'Cyclocity' scheme provides residents with 3,000 rental bikes, which between them have logged about 10 million miles since it started in May 2005. It has prevented an estimated 3,000 tonnes of carbon dioxide from being emitted, reduced vehicle traffic in the city by 4 per cent, and tripled bicycle use. In the Netherlands city of Houten, street layout gives pedestrians, cyclists and buses direct access to the city centre and railway station while limiting car access. As a result, cycling accounts for 44 per cent of all trips less than 7.5 kilometres, whilst walking accounts for 23 per cent.

As well as improving vehicle efficiency and making it easier to use alternatives to the car, emerging technology can also play a role in cutting the need to travel. British Telecom, for example, is pioneering the use of video conferencing technology. In a survey of teleconferencing by BT employees worldwide, 46 per cent of users said that their last conference call had saved them at least three hours of travel time. In total, BT saved an estimated £238 million in a single year through the use of teleconferencing, principally through avoided travel.

By encouraging businesses to implement a home-working strategy, they can radically reduce congestion and pollution. With the near ubiquity of broadband connections, and business servers that allow workers to log on as if they were in the office, travelling is unnecessary. For many years, executives have worked in this way; it's now time that everyone had the opportunity to improve their work/life balance, while also helping the environment.

Improving public transport to a standard where it offers a genuinely attractive alternative to the reliability and comfort of a private car is non-negotiable. Without it, we can't expect people to make the switch. Which is why our public transport can no longer lag behind Europe, while charging such high prices for its service.

Voter Demand Box:
Promoting the Alternatives

- **Introduce 'Cyclocities'**
- **Invest in buses**
- **Make dedicated school buses available for all schools**
- **Reward the use of technologies that eliminate the need to travel**

One of the reasons we've got it so wrong in Britain is that politicians have always tended to regard transport as an end, rather than a means to an end. Instead of designing policies to reduce the need to travel, politicians usually end up promoting developments which make it inevitable. What we should be asking is how can we get to our destination in a cleaner, greener way? And are there alternatives to forced travel? Through our planning system, the use of new technologies and our transport policies, we know we can promote smarter travel rather than simply more travel.

Step Eight

Built to Last

When we build, let us think that we build forever.
John Ruskin, 1849

There is a gulf between ordinary people and the modern architectural elite. That became strikingly clear in 2008 when English Heritage advised the British government not to block the redevelopment of Robin Hood Gardens, a run-down East London council estate. The advisory body described a 'bleakness of design' and said the estate was 'not fit for purpose'. And it was not alone. No fewer than 80 per cent of the people forced to occupy the buildings have said they want to be rehoused.

But despite this, the modern British architectural elite was enraged. One of Britain's most celebrated architects, Lord Rogers of Riverside, described the estate as a 'masterpiece. . . as good, if not better' than any other modern building in Britain, and even compared it favourably to Bath's great Georgian crescents.

Too often, the experts seem to forget what it is they are charged with doing. A home shouldn't be regarded as a temporary, abstract object, observed in isolation from its environment. It forms part of a community, a contributor to the quality of life of inhabitants and neighbours. Architectural originality may be important, but as a quality, it is less important than a proper emphasis on attractiveness, sustainability, quality of life, or what is now referred to as walkability.

'Walkable' neighbourhoods

A well-designed neighbourhood gives people access to the things they need – shops, schools, hospitals, offices – without dependence on the car. A poorly planned neighbourhood is one that increases car dependence, and fosters isolation.

Unfortunately, much development in recent decades has fallen into the latter category. In the US, the problem is stark, best illustrated perhaps by the fact that the distance people drive has doubled from an average of 4,000 miles per person in 1970 to 8,100 today. In Britain too, much of the growth of our towns and cities, and most of the new settlements we've built, have been constructed on the assumption of the car.

'When you begin designing transportation networks around people instead of cars, a whole set of good things happen,' says John Norquist, president and CEO of the US Congress for New Urbanism (CNU), a leading non-profit group promoting walkable neighbourhoods as an alternative to sprawl:

Throughout history, streets were expected to be vibrant public spaces and the setting for diverse and valuable economic activity, as well as movers of people and goods. When you design simply around vehicular movement as typically happens, you limit the results to a familiar landscape that includes big-box and strip retail. And perversely enough, you get a lot of traffic congestion and outrageous carbon emissions. It shouldn't be a surprise – when people need a car or truck to get anywhere, that creates a lot of long car trips.

President Obama made the same point. 'Over the longer term,' he has said, 'we know that the amount of fuel we will use is directly related to our land-use decisions and development patterns, much of which have been organized around the principle of cheap gasoline.' Changing this will take considerable time, but if we begin now by deciding what sort of future we want, the planning process will begin to deliver.

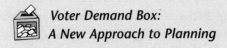

Voter Demand Box:
A New Approach to Planning

At the heart of all planning decisions should be an understanding of and a commitment to 'walkability'. New settlements and developments should be built wherever possible on the human scale, to promote local economic activity and with car dependence designed out of the system. Where plans are controversial, referendums proposed in step two should give people a veto where neccessary.

Where NOT to build

When politicians are faced with housing shortages, the reaction is always the same: promise to build more homes. The government's latest announcement, shortly before the onset of recession, was a plan to build 3 million new homes by 2020, mostly in the south and south-east of England. The goal was to reduce house prices by increasing supply. But the evidence has always been that other factors have caused high house prices in sought-after areas of England. Cheap loans and mortgages – now perhaps consigned to history after the sub-prime mortgage market collapse – fuelled house prices more than any lack of housing ever could.

Britain is a small nation, but unlike most other countries, we have allowed a disproportionate amount of activity to become centralized in and around one city – London. The effect is that pressure for housing in the south-east is immense, while other parts of the country are experiencing the emergence of ghost towns. The Empty Homes Agency estimated that in 2006 there were 633,328 empty homes – 3.6 per cent of the total in England – with 126,416 in the north-west alone.

It doesn't have to be this way. If, instead of promising millions of new homes in areas that cannot take them, the government sought to build better transport links across the country, businesses would inevitably spring up elsewhere rather than converge on London. If we built reliable and effective links between our cities, the incentive for people and businesses to begin repopulating parts other than the south-east would be significant.

Remote working could also radically change our approach to where we live and where we work. There is no reason why certain jobs could not be done from home or remote offices where workers could log in and begin work wherever they are. Teleconferencing could allow entire areas of the country to be revitalized simply because people would no longer feel the need to commute into the capital.

The government's chosen route is highly destructive. For instance, pressure on the south of England has led to an epidemic of so-called garden grabbing. Ministers say they don't keep records of the number of gardens that are sold off for development, but a recent survey of six London boroughs found that two thirds of all brownfield developments were in fact garden developments. And the government is explicitly culpable for this short-sighted approach, not least because it defines gardens as 'brownfield sites', meaning that instead of being protected as havens for wildlife, not to mention the sheer pleasure they bring communities, they currently have the lowly status of neglected industrial wasteland. Campaigners believe we are losing up to 30,000 gardens a year this way, the equivalent of twice the area of London's Hyde Park.

As we lose these gardens, paving over them with concrete and erecting more houses or office space, we are also at risk from increased flooding. When we build on grassland, gardens, meadows and other porous, natural surfaces with asphalt and concrete, rainwater has nowhere to run, it builds up and begins to flood our cities, towns and villages.

Our green spaces are precious, and the government should protect them. Even though it has far more land per person than we do, the US tax system is heavily biased against developing greenfield sites, and in favour of building on genuine brownfield sites. Any environmental clean-up costs, for instance, are fully deductible from income tax. The opposite is true in the UK. Our tax system makes it much easier to demolish old buildings and build new ones than to refurbish existing properties. If you want to build a new home, you pay zero VAT. If you want to make better use of an existing building, you pay the full rate. It's also easier to build on greenfield land than to clean up and develop brownfield sites.

We should follow the US lead. Gardens need the status of protected 'greenfield' sites. And, if there are over half a million empty homes in the UK, why not devise tax incentives to fill them? Why not introduce tax relief for people who want to let rooms in their homes to encourage better use of existing buildings?

It doesn't stop with gardens. Playing fields are similarly under siege. Although reliable figures are hard to come by, it is estimated that by 2002, school playing fields were being sold off at a rate of nearly one a week. Similarly under threat is the much-celebrated 'green belt'. Considered one of the nation's great post-war achievements, the fourteen green belts around England protect 1.671 million hectares of land – 13 per cent of England. But over 800 hectares a year are being destroyed, and 4,200 hectares are at risk from developers because the government is not willing to protect them unequivocally.

Worse still, despite the horrors of the floods in 2007, and promises by ministers that they would stop building on floodplains, the process continues unabated. The Countryside Alliance recently discovered that in 209 local authorities, nearly 150,000 new homes are to be built in high-risk areas.

Around Britain, 1.8 million households are at risk of flooding. And this is likely to get worse. For one thing, the population of England and Wales is set to increase by over 7 million by 2031. But more importantly, we continue to build in the wrong places. In the south-east, it is estimated that 30 per cent of new homes are to be built on floodplains. We even build in areas where we know the insurance companies won't provide cover.

In Aylesbury, Buckinghamshire, an area designated by the government for 'growth', 11 per cent of new homes will be built on land for which the Association of British Insurers will not guarantee affordable insurance. Selby in Yorkshire was one of the worst flooded areas in the UK in November 2006 – yet 3,000 new homes are due to be built there. The Environment Agency has warned that the cost of protecting London and the south-east from future flooding could exceed £20 billion.

We are building serious problems into the system, and the price will be paid by millions of people.

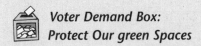

Voter Demand Box:
Protect Our green Spaces

High-speed rail to reduce pressure on the south of England

With massive investment in and creation of more fast rail links between our city centres – as described in Step Seven – we could realistically expect more businesses to take advantage of the lower cost of housing and offices throughout the UK. That way the vacant homes in the north would soon be filled, taking pressure off the country-side in the south.

Unequivocal protection for the green belt

The fourteen green belts around England protect 1.671 million hectares of land – 13 per cent of England. But over 800 hectares a year are being destroyed, and 4,200 hectares are at risk from developers. We need an unequivocal guarantee from our government that they will be protected.

Protect our gardens

It is absurd that under current rules, gardens are designated as 'brownfield sites' – making it easy for developers to build on them and for homeowners to sell them off for a quick profit. Urban gardens urgently need to be reclassified as greenfield sites, in order to protect them from destructive development.

Stop building on floodplains

The temptation is to call for higher flood barriers, more 'bunds' around rivers, barrages and engineered solutions.

But often these either simply delay the inevitable or actually make things worse. No one benefits from living on a submerged floodplain. We should demand of our government that it put in place an immediate ban on new development on floodplains anywhere in Britain.

Tax incentives for building in the right place
If you want to build a new home, you pay zero VAT. If you want to make better use of an existing building, you pay the full 15 per cent. This should be reversed. VAT should also be introduced for new buildings on greenfield land, with the money raised helping the development of brownfield sites.

Less tax on home letting
Bringing in tax relief on letting rooms in one's own home would encourage those who might currently be under-occupying property to take in lodgers, thus helping to ease the pressure on the housing market.

Efficiency in our homes

Inevitably, most of the arguments and discussions regarding climate change and emissions centre on transport and energy. But that only tells part of the story. While not as obviously polluting, the built environment – our homes, offices, schools and other spaces – are just as important. How we build our homes, how they relate to one another and where they are constructed all have huge social and environmental implications.

Around half of Britain's greenhouse gas emissions come from energy use in buildings. Domestic buildings are responsible for an estimated 27 per cent of that, with a further 10 per cent coming from the manufacture of construction materials.

As three quarters of the estimated 25 million homes in the UK will still be here in 2050, we need to look at ways to improve efficiencies in our existing homes. Up to 9 million houses have cavity walls but with no insulation. Millions lack even basic insulation or effective window glazing.

It's an easy problem to remedy. But there is a cost. In the medium term, improvements will save money on energy bills, and the cost of the upgrade will quickly be recouped. In the case of cavity wall insulation, the payback is about three years. But the initial investment, combined with the hassle, means people are reluctant to do the necessary work. It's for the government, therefore, to create genuinely attractive incentives, which will give homeowners the impetus to upgrade their home. Given that the average home changes hands every seven or so years, any system encouraging people to upgrade their homes at that point of exchange would theoretically deliver a clean and efficient housing stock relatively quickly.

Losing heat through walls is one thing – at least we know where it goes – with energy, however, it's more complicated. We don't know precisely where we are using it or in what quantities. Which is why so-called 'smart meters' are so important. If our homes were fitted with easy-to-read displays of their energy use and its cost, we would be able to monitor the effectiveness of our efficiency measures and see how using certain appliances consumes energy.

Wherever it has been introduced, smart metering has reduced household bills and cut emissions. In Sweden their introduction has led to a 20 per cent decline in energy use, while in the US the figure is around 27 per cent. Smart meters also make it easier for homeowners to sell home-grown energy back to the grid, as they allow for more accurate measurement of the energy generated, and its true value.

But no matter how we measure the energy we use, it makes

no difference if our household appliances are inefficient – which many of them are. Standby switches, for instance, account for 2.25 per cent of all electricity production, a figure that is set to increase as they are introduced into an ever-widening range of electronic equipment. And if every one of the country's 25 million mobile-phone chargers were left plugged in and switched on they would consume enough electricity to power 66,000 homes for a year.

Politicians often call on the public to make changes to their lifestyles, to buy better appliances. We've seen this over and over again. But when they engage in this kind of appeal, they are shirking their own responsibilities. They could, without any real difficulties, ensure a raising of standards across the board. They wouldn't have to require manufacturers to meet standards higher than what is already possible. It would mean taking best practice and rolling it into the norm.

A New Approach

Through retrofitting, the use of new technology and higher building standards and planning, it is estimated we could trigger a 60 per cent reduction in carbon dioxide from the housing sector by 2050. Government figures suggest that saving 30 per cent of the energy we use through cost-effective measures would in turn save the country £12 billion every year. If that's the case, why aren't they doing it?

Follow the best examples

The Hockerton Housing Project is a self-sufficient ecological housing development in Nottinghamshire. Its five houses use 75 per cent less energy than conventional homes. The houses are positioned in a way that allows maximum use of solar panels in winter, and they require no space heating at all. Not only that, but they harvest their own water and recycle waste materials causing no pollution or carbon dioxide emissions. It's one of the most ecologically impressive developments in the whole of Europe.

Further afield a ten-storey office building in Melbourne, Australia has reduced greenhouse gas emissions by 87 per cent, electricity consumption by 82 per cent, gas usage by 87 per cent and water use by 72 per cent. The building exterior moves with the sun to reflect and collect heat, and includes photovoltaic cells, solar water heating and rooftop wind turbines. It also turns sewage into usable water. The sustainability features added 20 per cent to the total cost, but payback is less than ten years.

Which begs the question, if we have the technology – and it works – why are these examples the exception, and not the norm?

The government is aware of our low standards, but its response has been to introduce ever more strident regulations. Despite fine intentions, it simply hasn't worked. The UK has some of the most cumbersome building regulations in Europe, yet still our standards are far lower than in other countries. There are a variety of reasons for this, but the main problem is the government's tendency to micro manage.

Cleaning Up the Government Estate

Central government-run buildings occupy about 17 million square metres of building floor space with over 5,000 km2 of land; some 2 per cent of the UK's land mass. It has been estimated by the National Audit Office that savings in energy and water consumption across the public sector could be at least £20 million a year.

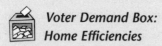

Voter Demand Box:
Home Efficiencies

Make it easier to upgrade our homes

The government should introduce a rebate on stamp duty for houses in which energy-efficiency improvements have been adopted at the time of sale, or within a reasonable time afterwards. At a stroke, this would provide all of us with a major incentive to pass on a much greener house to the next owner.

An alternative is to give all householders access to around £8,000 to invest in energy-saving technology with a payback period of five years or less. The money would be recovered automatically through the household energy bill over a 25-year period. This effectively means zero cost to homeowners who would see virtually immediate savings on their energy bills.

Work with the banks

Given that existing technology is already cost effective, energy efficiency theoretically already offers an attractive investment opportunity. Using well known landmarks like the Empire State Building, the Clinton Initiative for instance has shown that investments can be repaid within three or four years, after which there are considerable savings. The best long-term plan would therefore involve private institutions developing attractive investment opportunities in this field. UK pension funds for instance are sitting on about £860 billion. If they could be persuaded to invest a portion of that in energy efficiency initiatives, they would see a good return on the back of low-risk investments, and potentially huge sums of money would flow into low carbon initiatives. To provide reassurance and to trigger a rapid shift, the Government should underwrite initial investments. The only risk is that the price of energy may decline, which would simply lengthen payback times. But the opposite is more likely.

Smart meters in every home

There should be an immediate universal rollout of smart meters across the nation. If government pursued this policy energetically, 90 per cent of our homes could be fitted within five years. The estimated payback would be anything from four to six years, making it a good investment.

Better quality appliances

Government needs to set stringent minimum standards for energy efficiency. The standards should never be set beyond what is already possible, but they should be set high, raised often, and stringently enforced. If manufacturers are given notice that standards are to rise they will always manage to meet them.

Replace building regulations with standards

We need the government to abolish building regulations and replace them with new standards. These would be based on a range of outcomes: low emissions, efficient energy use, and the use of sustainable materials. It would be results that mattered, and this would provide an enormous impetus to finding new and more cost-effective means to deliver those outcomes. The government needs to trust builders to do the right thing, but be willing to impose heavy penalties where that trust is abused. If the regulatory system were turned on its head in this way, we'd see an explosion in innovation and a leap forward in standards.

As it stands, too many regulations clog the system. In Germany, builders are required to reach high standards of energy efficiency but they are not instructed as to how to do it. We too need to move towards a system based on outcomes, not processes. The government needs to trust builders to do the right thing, but be willing to impose heavy penalties when that trust is abused. If the regulatory system were turned on its head in this way, we'd see an explosion in innovation and a leap forward in standards.

Housing, perhaps like no other issue, affects us all, every day. The government's mass house building target has brought into focus the need for us to change the way we do things. If we fail – if we go on as before – we will see more of the countryside destroyed to build unnecessary houses that destroy the environment locally as well as globally. If we succeed, we will foster healthy communities, and set an example to the world.

Slow Water

The British used to understand their weather. It was unpredictable, but not extreme. Commuters fussed over whether they should take an umbrella to work, schoolchildren never knew when games would be rained off, a day that started like May could end like November – but whatever the disappointments and occasional joys, it generally didn't matter very much. In contrast to many other parts of the world, where water shortages or extreme weather events were commonplace, in these temperate islands we tended to take our weather – and our water – for granted.

Not any more. Our weather is changing. In 2006 we experienced some of the most extreme droughts and heatwaves for generations; a year later we were in the middle of some of the most horrendous floods for half a century. Whether or not global climate change is directly to blame for such specific events is impossible to say. But there is no doubt that our weather patterns are becoming more volatile. Life may become much harder for many people as a result. And – crucially – drinking water may become more scarce.

The increased likelihood of drought is only one reason for this. Over the past few decades, we have acted as if water was an infinite resource. We have wasted it in the household without a thought. We have built on floodplains, paved our gardens, polluted our watercourses with pesticides and industrial chemicals. Now we find that our carelessness has, paradoxically, both made flooding more likely at times of heavy rain and increased the likelihood of drought when heatwaves arrive. We have mismanaged our most vital resource, and now we are paying the price.

We need a fundamentally new approach – one that will remove the seesaw between drought and flood, which are part of the same destabilised cycle. We need to treat our water with respect and intelligence.

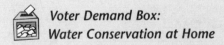 *Voter Demand Box:*
Water Conservation at Home

Over and above urgent measures outlined elsewhere to stop building on floodplains, paving over our gardens, and to reward farmers for looking after water meadows, we need to raise standards in the home.

Water-efficiency standards
The technology already exists for huge savings. German households, for instance, use a third less water than we

consume in the UK. Dual-flush lavatories and low-flow taps and showerheads can reduce household water use by up to 50 per cent, and such a reduction would also help to reduce the 24 per cent of domestic energy consumed in heating water. The government needs to set out minimum efficiency standards for water-using appliances, just as it should for electrical appliances. Standards should be strict but realistic, and under constant upward pressure.

Smart water meters

With smart meters in place in every one of our households, water companies would be able to implement a tariff that would provide sufficient water for families at a standard price, and then charge significantly more for consumption above that. Water metering usually leads to 10-15 per cent reductions in water use, and more at peak times. As with smart electricity meters, government should roll out smart water meters for every household in Britain as a matter of priority.

Step Nine

A Zero Waste Economy

Most of us are still too sane to piss in our own cistern, but
we allow others to do so and we reward them for it. We
reward them so well, in fact, that those who piss in our
cistern are wealthier than the rest of us.

Wendell Berry

Ten years ago, American yachtsman Charles Moore decided to
take a short cut through the North Pacific Gyre during a yacht
race between Los Angeles and Hawaii. It is an area of the
Pacific that sailors try to avoid because of the slow and
complicated winds that prevail. What he discovered was
enough to persuade him to give up racing and become a full-
time campaigner. Thousands of miles from land, Moore found
himself surrounded by seemingly endless 'plastic soup'. 'Every
time I came on deck, there was trash,' he said. 'How could we
have fouled such a huge area? How could this go on for a
week?'

Unknown to him at the time, he had stumbled across the
world's biggest rubbish dump, roughly twice the size of Texas;
a vast mass that stretches from about 500 miles off the
Californian coast, across the Pacific almost as far as Japan. It
is the single largest body of pollution anywhere in the world,
and is believed to contain about a hundred million tonnes of

plastic, kept together by swirling underwater currents. It is, effectively, a giant plastic continent. About one fifth of the rubbish is thrown from ships, and includes mile upon mile of plastic cord and disused fishing nets. The rest comes from land.

Among the waste, there is a substantial quantity of plastic pellets, which absorb surrounding chemicals and enter the food chain. 'Whether it be an algae-sifting whale or a fish-eating seal, small pieces of plastic are mistaken for food at all levels of the chain,' says Moore. At the top of the food chain, much of that contamination ends up in our own bodies.

If this trend continues, the plastic 'soup' will double in just ten years. There it will stay for decades, in some cases centuries. Roughly 70 per cent, according to the UN, will eventually sink to the bottom of the ocean, where it will join what has already become a mountain of toxic waste. The plastic is causing ecological mayhem. Charles Moore has counted more than 100,000 Laysan Albatross deaths in a single year and believes the species is heading for extinction, along with the Hawaiian monk seal and many other species.

Sadly, this is not just the opinion of one man. The UN Environment Programme has said that every square mile of ocean contains 46,000 pieces of floating plastic, which it has linked to the death of more than a million seabirds every year, as well as more than 100,000 marine mammals. The Pacific plastic soup is the starkest illustration of our throw away global economy – one in which Britain plays a leading role.

Every hour, the UK throws away enough rubbish to fill the Albert Hall to the ceiling. That's 335 million tonnes of waste a year – an incomprehensibly large figure, which causes groundwater pollution, resource depletion, and contributes to climate change at every stage of the process. In England alone, roughly 30 million tonnes of construction and demolition material is sent to landfill as waste, and more than 6 million tonnes of household food waste is thrown away annually.

The vast majority of this waste ends up in the country's rapidly filling landfill sites. Packed into these gaping eyesores, our rotting rubbish releases noxious fumes and vast quantities of methane – the most destructive greenhouse gas, worse by far than carbon dioxide. When not dumped into the earth, our rubbish is piled into deeply unpopular incinerators and burned instead.

We bury it, we burn it – and only after we have exhausted these options do we start to think about recycling. Just 27 per cent of our domestic waste is recycled. Compare this with the 70 per cent in Holland, or the consistently higher levels across most of Europe. The irony is that opponents of environmental politics will cite cost as their principal concern. But the profligate way we generate waste is an example of precisely the opposite; the national bill for disposal of this waste is growing by the year – landfill costs alone have reached the £10 billion mark.

The government recognizes that we need a culture change. 'If every country consumed natural resources at the rate the UK does,' it reported in a recent waste strategy document, 'we would need three planets to live on.' And yet we are more wasteful than ever. Most resources stay in the economy for less than six months before being discarded. Paul Hawken, author of *Natural Capitalism*, states that 94 per cent of the materials used in the manufacture of durable products become waste before the product is even ready for market. Clearly waste is not just a British problem – the European Environment Agency says that by 2020 enough household rubbish will be being produced in Europe each year to cover the island of Malta to a depth of seven and a half feet.

We are running out of landfill sites. Incinerators are

polluting and unpopular. Raw materials are becoming costlier, while the value of the material we could be recycling, rather than destroying, will reach £1.8 billion per year by 2020. Yet progress remains resolutely sluggish. In 2008, the British government teamed up with other EU nations to block proposals to establish binding targets to stabilize waste at current levels by 2012.

These targets are vital, not because they will in themselves bring about change, but because they provide a yardstick against which the government can be judged. To actually cut back on waste requires us to think very differently. In the long term we should aspire to become a zero-waste economy.

Zero waste and magic carpets

Britain does not have to look far to see some of the world's most successful waste-reduction schemes. Europe is leagues ahead of us at developing systems that can really make a difference. The Dutch now recycle 70 per cent of their batteries. The Swiss have boosted paper recycling from 50 per cent in 1990 to 70 per cent today, as well as recycling 96 per cent of glass containers.

In Denmark, recycling is free and use of landfill very expensive. As a result landfill has been cut by about a quarter. Just 3 per cent of Copenhagen's waste goes to landfill. This was achieved through a combination of schemes including the establishment of furniture recycling centres, composting, and the consumer right to return unused products, like paint, to retailers. All bottles, paper, newsprint, cans, cardboard and 85 per cent of building waste go to recycling. The city saves £7.7 million annually. Similarly, Flanders, Belgium, has one of the highest recycling rates in Europe, at 71 per cent.

Further afield, Canberra, Australia's capital, has a target to eliminate waste by 2010. In practice this means 95 per cent of waste will be recycled and just 5 per cent will go to landfill, with no incineration. The city is currently recycling 73 per cent of all household and commercial waste. The amount of waste going to landfill fell by 42 per cent between 2000 and 2005. In 2001, New Zealand set itself the goal of becoming the world's first zero-waste nation by 2020. Such strident ambitions should be the benchmark to which we aspire.

To accomplish this, however, requires a genuine willingness to make a change – not just at consumer level, but also within business. Fortunately there are extraordinary examples to follow.

Take the giant carpet company, Interface. Modern carpeting is hugely wasteful. It lasts on the floor for an average of twelve years, and then spends 20,000 or more years in land-fill. In 1996 the company's director, Ray Anderson asked his staff to work out his company's ecological footprint. He was staggered to learn that 1.2 billion lbs of raw materials had been extracted to produce the £485 million worth of carpets sold by the company the year before.

Shocked by this incredible figure, Anderson and his company decided to rethink their business model. Instead of simply selling carpet, the company now sells a carpet service. Customers pay a monthly fee for a service that guarantees them permanently fresh-looking carpets. As the carpet tiles wear out, they are replaced by the company and recycled. The effect is that clients always have good quality carpets, and the company has a clear incentive to provide carpets that last.

Since the initiative was introduced in 1995, the company has diverted a mass of more than 100 million lbs worth of material from landfill. The energy used to produce the carpets is down 41 per cent, the equivalent of 61,000 barrels of oil. Emissions reductions are down 56 per cent, the equivalent of taking 21,000 cars off the road for a year. Water use is down 73 per cent. And crucially, the initiative has saved the company £192 million through eliminating waste. At the same time its sales have increased to £606 million. Meanwhile, Interface has expanded to become the world's largest seller of modular carpet tiles.

Xerox, the document management company, has also broken new ground, with its 'Xero Waste' initiative. Even as far back as 1998, the company achieved a worldwide recycling rate of 88 per cent, saving itself £27 million. It wants to go further. The company developed the first photocopier that is 95 per cent recyclable/reusable. When the photocopier reaches the end of its useful life, Xerox collects it, dismantles it into its component parts, and rebuilds it into another machine. In the last ten years Xerox's 'remanufacturing' has given new life to the equivalent of more than 2.6 million copiers, printers and multifunction systems. Of the 37,000 metric tonnes collected in 2005, Xerox reused or recycled 98 per cent. It has diverted 107 million lbs of material from landfills through equipment remanufacturing and 14 million lbs through reuse/recycle of supplies.

How do we create a zero-waste society?

Policymakers need to understand that consumers don't welcome waste. But living in a wasteful, throwaway economy makes it almost impossible for individuals to cut back. We could start by dropping the term 'consumer waste'. It's nothing of the sort. Consumers don't demand excess packaging – producers provide it.

If people had the right to take any packaging waste back to the shop it was bought from, retailers would be obliged to deal with that waste directly. If the chains are required by law to deal with the mess, they'll design their products more responsibly in the first place – or pass on the pressure to those who do. Similar approaches work elsewhere.

For the remaining waste, households should be encouraged to recycle. A new company in the US, Recyclebank, offers householders considerable financial rewards for minimizing waste and increasing recycling. It has had extraordinary results, boosting recycling in some cities from a meagre 9 per cent to 40 per cent. There is no downside, but the upside is significant. Instead of paying cash, the company awards coupons that can then be used in participating shops. This and similar schemes need to be introduced in the UK – preferably with the coupons attached helping smaller independent retailers that need all the support they can get.

But this can only go so far. Household waste still only represents a tenth of total waste; the remainder is generated by business. The simple, blunt and most effective way of minimizing this waste is to make it financially unattractive to produce it in the first place – which is why a meaningful landfill tax would be a hugely effective instrument for change.

If waste becomes a financial drain, companies will respond by cutting the amount of waste they generate. Companies like Uponor have already demonstrated that. One of the country's biggest construction firms, it has cut the amount of rubbish it sends to landfill by 86 per cent in ten years, and has saved a great deal of money in the process. Given that the construction sector is responsible for three times more waste than all the households in the UK combined, Uponor provides a hugely valuable example of best practice.

Landfill tax has existed since 1996, but the level of tax has remained virtually unchanged and to be truly effective, needs to be greatly increased, with a large percentage of the money raised going directly to the communities and landscapes affected by nearby landfills. As landfill tax rises, good companies enjoy increased comparative advantage. Uponor, for instance, would do even better than wasteful competitors. In addition to an increase in the level of the tax, landfill should no longer be an option for any waste that can be recycled, composted or reused.

But there is more we can do to encourage companies to make products that last. Tough standards need to be set to ensure that manufacturers waste as little as possible – and that it is in their economic interest to conserve resources and recycle waste. For example, we could require all drinks cans to be made from 100 per cent recycled aluminium, or 100 per cent of all computers to be recycled by the manufacturer.

In 2007, a staggering 160 million computers across the world were thrown away. If the manufacturers adopted a modular design, where every component of the machine is a separate pluggable element, instead of being simply thrown away, computers could be dismantled and reused. In Japan, all television sets, air-conditioners, washing machines and refrigerators must be between 50 and 60 per cent recyclable. Brand new computers now have fewer parts, which has also saved on breakdown costs. Since this law came into effect in 2001, the number of refurbished computers sold back to consumers has hit a million a year.

A tax could also be introduced on the use of virgin materials to stimulate the markets in recycled equivalents. If extended to the use of non-recyclable materials in new products it would also discourage the use of materials that can only be used once. Whatever money is raised should be reinvested, along with part of the proceeds of landfill tax and others proposed here, in paying for the infrastructure we will need if we're to achieve a zero-waste economy.

An example of how this money could be invested is Vienna's Repair and Service Centre (RUSZ). It combines reusing and recycling waste electrical goods that would otherwise end up in landfill or incinerators with the reintegration of long-term unemployed people. So far it has been able to prevent around 2,000 tonnes of waste per year. The RUSZ has kick-started the development of an entire network of repair enterprises in the city.

So often with consumer goods, it is cheaper to buy a new one than to have it repaired. Things aren't built to last. The economy is flooded with temporary and disposable consumer goods. They may be cheap to buy, but they don't last, and when they break down, they accumulate in landfill sites across the country. We need to find a way to encourage manufacturers to stop making things with built-in obsolescence.

One solution is to introduce a levy on goods with an average life expectancy of 20 per cent less than the product average. So if toasters, on average, last ten years, a toaster with an average lifespan of eight years or less would be taxed, with the money, as ever, being used to bring down the price of those other products designed to last, or be readily disassembled and reused.

From Bath in the UK to Canberra in Australia, and even entire countries such as New Zealand, people have declared their ambition to create zero-waste communities. It's only an aspiration, but it is one Britain as a whole must embrace. It goes without saying that we cannot hope to achieve a constant, sustainable economy, unless we design waste out of the way we do business and live within our means.

Voter Demand Box:
Blueprint for a Zero-Waste Economy

'Take Back'

People should have a legal 'take back' right enshrined in consumer law. This would give everyone the right to take any packaging waste back to the shop it was bought from, and impose an obligation on retailers to recycle that waste once it was received.

Paying people to recycle

Where householders can earn money by recycling, and save money by generating less rubbish, the recycling rates soar.

No more landfill

In the shortest possible time, landfill should be banned for anything that can be recycled, composted or reused instead. Landfill tax for remaining rubbish needs to rise, with the proceeds being re-invested in communities affected by landfill sites, and in infrastructure for further reducing waste. The government is planning to raise the tax to £48 per tonne by 2010, but to make a difference the increase needs to be much bolder.

Using the right materials

A tax should be introduced on the use of virgin materials to stimulate the markets and innovation of recycled equivalents. The tax should be extended to the use of non-recyclable materials in new products to discourage the use of materials that can only be used once. The proceeds should be used to bring down the price of recycled materials.

Built to last

Introduce a life-expectancy levy for goods that have an average life expectancy of 20 per cent less than the product average. The proceeds should be used to reduce the price of long-life alternatives.

Government buying power

The government must use its vast buying power to stimulate the market and drive innovation. Across the board, its procurement contracts need to have an in-built bias in favour of recycled and sustainably sourced materials.

Incineration, a last resort

New technology probably makes the process less hazardous, but there is still the problem that contracts with incinerator operators tend to displace activities around recycling. Incineration should be a last resort, of the highest possible standard, and only ever allowed where local people approve it, if necessary via a referendum. Incinerators should be built to generate power and capture heat – in other words, to work as part of a combined heat and power system.

Step Ten

Playing Our Part

Look to your consciences and remember that the theatre of
the world is wider than the realm of England.
Mary, Queen of Scots, 1586

Most environmental problems, by their very nature, are international. From climate change to overfishing, water shortages to species loss, the world's most pressing issues ignore national borders. As a consequence, those sceptical of environmental action point out that the UK produces just 2 per cent of global emissions. What difference can our little country make to climate change, they say, when compared to the huge carbon emissions of the US, China and India?

If we're going to rate ourselves in terms of our standing in the world's pollution league, we need to get the numbers straight. The 2 per cent figure is misleading. Include the global impact of investments made via the City of London, for example, and that number swells dramatically. Add in the impact of imports and the figure increases again. Then take into account our share of global aviation emissions (6.1 per cent of all international air emissions) as well as the contribution of the products sold by the five largest UK oil and mining companies, which account for 10 per cent of all global greenhouse gases, and our real contribution to the problem is much bigger than we like to think.

Arguing over percentage points is, however, pointless. Whatever our precise percentage contribution to the ecological crisis is, we have a duty to act to improve the situation. We are one of the richest countries in the world, and one of the planet's biggest economies. If we cannot green our way of life, how can poorer countries be expected to? And why should they listen to us when we talk about it? We need to take a lead.

There is a practical reason, too, for international action. If we are to shift from a dirty, wasteful and polluting economy towards a system that understands the value of the natural world, it will be through changing the framework in which the market operates. It will involve finding ways to price the environment into the market. It will involve shifting the tax burden away from good things like employment towards bad things like pollution and the misuse of scarce resources. It will involve putting a price on carbon, raising standards for cars, electrical appliances and biofuels, and banning the worst offenders like hydrofluorocarbons (HFCs).

Virtually none of this activity can realistically be confined to a single nation. If, for instance, Britain had decided, alone, to ban chlorofluorocarbons (CFCs), the environmental benefits would have been minimal. It is because world governments signed up to the Montreal Protocol in 1989 that 90 per cent of the global production of ozone-depleting substances like CFCs has been phased out, and the rate of ozone-layer depletion in the stratosphere is now declining. We are seeing the first signs of the recovery of the ozone layer.

However, that progress risks being undermined by the continued use of so-called super greenhouse gases such as HFCs, which are used in fridge manufacture. HFCs' effects are between 100 and 3,000 times that of carbon dioxide. An aggressive phase-out schedule for HFCs is technologically and economically feasible, not to mention utterly essential, but it won't happen unless ratified on an international level.

International agreements are, however, only as good as the targets they set, the countries that sign up for them, and the adherence they show to them once they have put pen to paper. Agreements at Kyoto in 1997 and in Bali ten years later were inadequate. At Kyoto, little was agreed, and much of what was promised never happened. At Bali, no targets were set to reduce emissions. However, the global mood is clearly changing, and there is a real opportunity now to press for a coherent global agreement to cut emissions by a minimum of 80 per cent by 2050. According to the science, that offers the best opportunity to keep a global temperature rise below 2 degrees above present levels.

How to go about actually reducing those emissions is a contentious issue. But it seems quite clear that there are few better options than carbon pricing. Without it, companies have very little pressure to clean up. Consumers and businesses need to see the environmental and economic costs of carbon emissions translated into actual financial liabilities. Clear price signalling will make polluting activities more expensive and cleaner activities less expensive.

The European Union Emissions Trading Scheme (EUETS) is the largest 'cap and trade' system in the world. It works like this: EU governments agree a cap on emissions for different sectors of the economy within each EU country. Carbon quotas are then allocated by those countries to individual businesses. The quotas are tradeable, so companies that pollute less than they are permitted to can sell their excess quotas to those who need them. Emissions trading is effectively a transfer of wealth from polluters to non-polluters.

The first phase of the EUETS was a failure, principally because the national allocations were set far too high. We therefore have to get the next phase right. The allocations need to be realistic, and the permits need to be auctioned, not merely handed out to companies. If industries have to pay for their quotas, they will be far more likely to value and act on them. It's also important that more sectors are included in the EUETS. It currently covers 45 per cent of all emissions including power plants, steel, cement and paper manufacturing. But aviation, for instance, is excluded, along with manufacturers of aluminium and chemicals. They need to be included.

President Obama has announced his own intention to establish a US cap-and-trade programme. The initial allotments of carbon credits will ignite a national debate, and so too will the steepness of the proposed emissions reductions, but most US utility companies accept the need to price carbon, and the certainty this scheme would bring will allow them to make informed plans for the future.

Forest services

While progress on carbon pricing is slow, it is nevertheless seen to be happening. This is more than can be said for the protection of our forests. Despite 'saving the forests' being the rallying cry of the environmental movement more or less since its inception, progress – real, tangible help for these crucial natural resources – has been marginal at best.

The world's great tropical forests are one of the most remarkable features of our planet – critical for maintaining the health of the atmosphere, regulators of rainfall and fresh water around the world, and home to an estimated 50 per cent of the planet's species. They represent a giant, yet dramatically under-valued global utility. The world's forests, leaf litter and soil store roughly 50 per cent more carbon than the atmosphere.

Because of that, deforestation is responsible for approximately one fifth of global greenhouse gas emissions – about the same as the contribution made by China's economy.

Each day, the trees of the Amazon release 20 billion tonnes of water into the atmosphere – the energy equivalent of the world's largest dam operating at maximum power, every day for 145 years. On top of this, the world's forests are home to more than a billion people who depend on them for their survival. These extraordinary services remain undervalued, even unvalued by the markets, but their loss would wreak havoc on a global scale. And we are losing them. Tropical forests continue to be destroyed at an unprecedented rate in pursuit of short-term economic gain. It is one of the great tragedies of our age.

Brazil is South America's largest economy. It holds around one third of the world's remaining rainforests, including 60 per cent of the Amazon. Between 2000 and 2005, some 34,660 km2 were destroyed every year, adding to a total deforested area of 616,000 km2, twice the size of France, and representing 15 per cent of the total forested area. Much of this has happened since the seventies. If trends continue, 40 per cent of the Amazon will be destroyed by 2050, and on top of the loss of biodiversity and rainfall, more carbon will be released into the atmosphere than the entire world generates in three years, at 2000 levels. The trees of the Amazon rainforest are estimated to store 90–140 billion tonnes of carbon, and deforestation accounts for about 70 per cent of Brazil's greenhouse gas emissions.

To say this needs to stop is only to echo the calls of the early days of the environmental movement, but that doesn't make the message any less important. The international community has to find a way to ensure that standing forests are worth more than destroyed forests. It's not a new concept, and nor is it particularly radical. Indeed some of the forest nations are calling for action, knowing as they do that if they destroy their forests now, they will pay in the medium and long term. But they also know that unless they can generate income from looking after their forests, they will not be able to, or want to hold back the pressures of development.

In 2008, the president of Guyana, Bharrat Jagdeo, issued a challenge to the international community, and specifically to Gordon Brown, to help him save his forests in the fight against climate change. He has offered to put his country's entire standing forest under internationally verified protection in return for payments that will support a new model of sustainable development. The giant consultancy firm, McKinsey, has offered to help Guyana design the mechanism that will make the country's forests more valuable alive than dead, but so far neither Gordon Brown nor any other governments have responded to President Jagdeo's repeated requests for help.

'Now we need solutions that are national and supranational in scope, and which address the fundamental reality that deforestation is a result of a market failure that makes trees more valuable dead than alive,' the president said. 'In Guyana, we are ready to play our part, and to provide a model for other rainforest countries to share. Our deforestation rate is one of the lowest in the world and we want it to stay that way. However, we also face considerable development challenges – we need better schools and hospitals, more jobs and economic opportunities, and to meet all the other economic and social demands of Guyana's people. I frequently receive proposals from investors to convert our forest into land for agriculture or biofuels. Agreeing to these would be a quick way to meet the development

175

challenges we face – but in Guyana, we are acutely conscious of climate change.'

Brazil too has shifted from a near absolute refusal to engage with the international community on the issue of the Amazon, to a clear willingness to act. More than that, the country is crafting a radical programme to cut deforestation by 70 per cent by 2017. Brazil's government knows that it cannot hope to look after the forests without considering the livelihoods of the many millions of people engaged in its destruction. As a result, it plans to legalize the status of some 25 million squatters. By giving them ownership of the land they currently occupy, they will have for the first time a valuable asset, and a direct incentive to protect it. The scheme is designed to favour smallholders, who will not be required to buy their land.

The plan goes much further. Up to 16 million hectares of land that was once Amazonian forest is degraded and abandoned, an area the size of England and Wales combined. Throughout the country as a whole, there are an estimated 100 million hectares of abandoned land. It is abandoned because using it productively costs roughly 20 per cent more than the alternative of clearing virgin forest.

The government will therefore create a series of strong tax incentives for the regeneration of degraded land, so that it becomes less expensive to operate on it than it does on freshly cleared land. In addition, the government will make attractive financial packages available to farmers who want to engage in sustainable activities. All this allows the government to then pursue an aggressive campaign against illegal logging, which is a key cause of destruction.

Brazil has already put aside substantial resources for the project, but it needs much more, and for the first time is seeking to work with international partners to establish sufficient funds. It has set up an Amazon Fund and is inviting contributions from other governments. Norway has already committed £606 million over ten years.

The key for Brazil is control. There is no question of the country losing sovereignty over its land or resources. Contributing countries will be able to verify Brazil's performance, and it will be on that basis that they continue to provide support. If it works, the project could become a model for other forest nations. For the first time, then, there is good news, although the history of international efforts to halt deforestation is littered with failures. More often than not, the money has ended up in the wrong hands and forest clearance has continued much as before.

For example, the Tropical Forestry Action Plan (TFAP) of the eighties and nineties was widely criticized and failed to achieve its goal of sustainable forest management. This was because the TFAP did not attend to underlying issues such as land rights; it was non-transparent; it was seen to be donor-driven (rather than reflecting the priorities of the rainforest countries themselves); and more importantly it funded projects that made matters worse, not least providing finance for industrial logging operations.

For years, campaigns to save the forests have focused on biodiversity, indigenous people and the erosion of environmental services. In an ideal world, these assets would have a value sufficient to ensure their protection. The fact that they don't illustrates the crudeness of our ability to measure what matters. But they don't, and we do not have sufficient time to devise an entirely new mechanism. As Paul Hawken, author of *Natural Capitalism* said, 'While there may be no "right" way to value a forest or a river, there is a wrong way, which is to give it no value at all. How do we decide the value of a 700-year-old tree? We need only ask how much it would cost to make a new one. Or a new river, or even a new atmosphere.'

What is needed is a mechanism by which we pay now, rather than later. The questions are, where does the money come from, how do the forest nations access it and how should it be spent? Some countries favour the establishment of a global fund, raised via contributions from developed nations, perhaps using the proceeds of EUETS carbon permit auctions, and dipped into by forest nations on the basis of avoided deforestation.

In October 2008, the European Commission proposed earmarking 5 per cent of the auctioning revenue raised from the EUETS allowances for global efforts to combat deforestation. That would raise something in the region of £1.2 billion to £2 billion. However, some experts believe the challenge will cost more. Nicholas Stern believes we need £6 billion per year to cut tropical rainforest loss by 70 per cent.

If this is the case, a market mechanism would be needed to raise that kind of money. If avoided deforestation, which is currently excluded from the carbon markets, was included, the necessary finance would become more easily available. Using the carbon markets could raise sums of money that would dwarf the quantities likely to be made available through direct governmental contributions.

More or less everyone agrees that whatever the eventual mechanism, the process will inevitably require public funding in the short term.

Taking the lead on forests

There is a growing hunger for a global solution, and a growing consensus that it will involve developed countries compensating forest nations for preserving their forests, or paying for the services provided by the forests. The UK needs to take the lead in ensuring that the so-called United Nations REDD negotiations (Reducing Emissions from Deforestation and Forest Degradation) can deliver proper resolution.

But having established the principles, and the mechanism, brings a host of challenges. Simple laws of supply and demand will necessarily boost the cost of commodities. More people, greater consumption, diminishing productive land and increasing use of that land for biofuels, all lead to increased pressure to exploit forest lands. Whatever mechanism is developed to protect forests must ensure that the levels of funding remain greater than the potential profits involved in destroying forests. The compensation levels will therefore have to be linked in some way to an index of commodity prices.

The mechanism must also distinguish between plantations and natural forests. If it doesn't, we may see forests destroyed, the timber sold and the land replanted with artificial monoculture, all paid for by well-intentioned foreign governments. There are also concerns about land tenure issues in relation to carbon trading, and who will benefit from the trade in credits. The Brazilian approach, if it is followed through and if it works, will inspire others to seek solutions.

Brazil's earlier opposition to absorbing standing forests into the carbon markets was based on a fear that rich countries would use cheap forest carbon credits to trade their way out of historical responsibility for greenhouse emissions. As President Lula de Silva told a meeting of the G8 in June 2007, 'everyone knows that the rich countries are responsible for 60 per cent of the gas emissions, and therefore need to assume their responsibilities'. In the context of the battle against climate change, forest-derived carbon credits must not become an easy 'way out' for developed countries.

Placing a value on healthy forests is crucial. But at the same time, we need to stifle the market in illegal timber. Nearly a third of the timber and timber products imported into the EU is believed to be illegal. For the UK, WWF estimates the figure is roughly a quarter. Given that Britain is the world's fourth biggest importer of timber, that represents a big part of the problem.

New laws have been introduced in the US to ban the trade in illegal timber, and we need the same in the UK and across the EU. Perhaps not surprisingly it is the UK timber industry itself that is leading calls for tighter laws, to protect itself against cheap competition. Eighty-two timber companies and associations recently signed a statement calling on the EU to 'adopt new legislation which makes it illegal to import illegally sourced timber and wood products into the European market'. British signatories included B&Q, Sainsbury's, Travis Perkins and the Chartered Institute of Builders.

There's more that the government can do to support sustainable timber production, not least in government departments themselves. Public sector procurement accounts for about 40 per cent of all timber purchase in the UK.

If pollution and the use of scarce resources become a financial liability, we will see the very DNA of businesses begin to change, and a low-carbon economy will become an inevitability. Similarly, if we can place a value on the true services provided by the world's forests, they will survive and thrive. Instead of watching as they are transformed into disposable goods, loo roll or biofuels, we must act now to protect them. Britain may not be as big as it used to be, but it has long arms and great influence. It's time to use them.

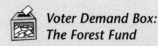
Voter Demand Box:
The Forest Fund

The UK should push for the establishment of an international 'Forest Fund' and work closely with forest nations to determine how best to fill it. There are three options; direct payments by individual countries, perhaps using the proceeds of the sales of carbon permits; direct use of the carbon markets by allowing the trade in forest-related carbon credits; or more likely, a combination of the two, with direct payments being made immediately while the carbon markets develop.

Voter Demand Box:
Sustainable Timber

The UK must seek to eliminate the trade in illegal timber, and push for radical change across the EU. It should introduce a no-compromise requirement on the public sector to source sustainable timber only.

Voter Demand Box:
Carbon Pricing

The first phase of the EUETS was a failure, principally because the national allocations were set far too high and there was no pressure to cut emissions. We therefore have to get the next phase of EUETS right. The allocations need to be realistic, and the permits need to be auctioned, not merely handed out to companies. If industries have to pay for their quotas, they will be far more likely to value and act on them. It's also crucial that more sectors are included in the EUETS. It currently covers 45 per cent of all emissions including power plants, steel, cement and paper manufacturing. But aviation, for instance is excluded, along with manufacturers of aluminium and chemicals. They need to be included.

Conclusion

*And God saw everything that He had made, and found it
very good. And He said, This is a beautiful world that I
have given you. Take good care of it; do not ruin it.*

Jewish prayer

In *The Man Who Mistook his Wife for a Hat*, eminent
psychologist Dr Oliver Sacks describes the case of Dr P, a
distinguished musician and teacher.

> I opened a copy of the *National Geographic* magazine,
> and asked him to describe some pictures in it. His
> responses here were very curious. His eyes would dart
> from one thing to another, picking up tiny features, indi-
> vidual features, as they had done with my face. A striking
> brightness, a colour, a shape would arrest his attention
> and elicit comment – but in no case did he get the scene
> as a whole. He failed to see the whole, seeing only the
> details, which he spotted like blips on a radar screen.

Later Dr P is presented with a bundle of photographs of his
family and close friends. 'It was not merely the cognition at
fault; there was something radically wrong with the whole
way he proceeded,' Sacks notes. 'For he approached these
faces – even those near and dear – as if they were abstract
puzzles or tests.'

And when Dr Sacks handed his patient a freshly cut red rose,

> he took it like a botanist or morphologist given a specimen, not like a person given a flower. 'About six inches in length', he commented. 'A convoluted red form with a linear green attachment.' And what did he think it was? 'Not easy to say.' He seemed perplexed. 'It lacks the simple symmetry of the platonic solids. . .' 'Smell it,' I suggested. He looked somewhat puzzled, as if I had asked him to smell a higher symmetry.

I was fascinated by Dr Sacks' book, and this case in particular, when I read it shortly after leaving school. Many years later, however, while rereading the book on a long train journey through Europe, it struck me that prosopagnosia, as Dr P's illness is called, may not be half as rare as the author first believed.

Indeed, Sacks might just as well have been describing society as a whole. With the parts, we perform miracles. Who can fail to be amazed and humbled by the skills of our surgeons, the magic of Google, the wonder of mobile telephones? We have advanced our understanding and our abilities in innumerable areas, but when it comes to solving systemic problems, there is a block. Like Dr P, we compartmentalize the world around us to such an extent that we only ever seem to understand a small portion of the picture.

Take cancer. Year-on-year instances of cancer have increased. It affects one in three of us, meaning that everyone is touched by it, directly or indirectly, at some stage in their lives. With every promise of a miracle cure, we breathe easier, and praise the advances in medical knowledge. But why do we not stamp our feet in anger that virtually none of the billions upon billions of pounds spent looking for cures is apportioned to finding the causes of this pernicious disease? Surely if the number of cancer victims is increasing, something is causing it to increase?

It's the same principle as the gruesome increase in early-onset puberty. If one in every hundred 3-year-old American girls is going through puberty, why is that fact met with indifference by much of the science community? How can it be right that we spend more money developing drugs to manage 'precocious puberty', and surgical skills to normalize sexual organs, than we do preventing these deeply worrying things from happening?

We pursue an extreme safety agenda in relation to our children that borders on the hysterical and is gradually depriving our children of their childhood. Conker fights, climbing in trees, school trips, even noisy school bells have been targeted by 'Health and Safety' regulations. But where is the protection when it is actually needed?

This, almost deliberate, rejection of common sense is widespread. When the Royal College of Psychiatrists was asked to look into the causes of farm suicides, they cited one reason: a higher level of shotgun ownership. Not that farm income had collapsed, or that the industry was in crisis, or that farmers were hideously undervalued. But that they had better access to one method of self-harm than other sectors of society. By the same logic, windows should be banned in tall buildings.

Unbelievably, these findings weren't greeted with hoots of derision. In our scientific age, we have a tendency to believe whatever those in white coats tell us is true, even when we know it is patently false. When Sir Richard Doll was asked to look into cancer clusters surrounding British nuclear power plants, for example, he didn't identify the obvious link with radiation. He claimed it was caused by a 'leukaemia virus', for which there wasn't, and still isn't, a shred of evidence. He remained until his death one of our most respected scientists.

But it is in the field of economics that this blindness has the greatest implications. The economy sees no value in nature until it is cashed in. It behaves as if it is nets, not healthy fisheries that provide us with fish, as if saws rather than forests provide timber. It tells us that commodities are valuable, but the systems that provide them are not.

Common sense tells us the opposite. Common sense says that it's better to combat the cause of a disease rather than wait for a cure. Common sense says that without the natural world there can be no economy. Common sense says that over-consumption will eventually lead to scarcity. But at an administrative level, common sense has been ignored. Dr P has been in charge. It's time for him to stand down.

At long last, there are signs of a shift in our thinking. Across the board – economists, religious leaders, politicians, company executives – are talking a new language. Finally, there is hope. We now have an opportunity both to avoid catastrophic change and to find a new way to flourish as a species.

But we have to get it right. The danger is that we invest disproportionate hope in technology alone. There is zero doubt that without a revolution in clean technology and the emergence of an age of efficiency we will never be able to live within our means. Yet we also need to recognize that there is something wrong in our approach. No matter how advanced a technology, we will never be able to renew biological systems. We could never create an ecosystem, and if we could, we couldn't afford it.

It has been tried. In 1991, £121 million was invested in the so-called Biosphere II project in Arizona. The goal was to create a self-contained, self-regulating ecosystem in a three-and-a-half acre enclosed dome designed to sustain eight people for twenty-four months. The experiment was a disaster. The enclosed air soon became unbreathable, with a diminishing supply of oxygen and an increase in nitrous oxide. Over three quarters of the dome's small animals died out. The trees were outcompeted by opportunistic vines. The insect pollinators died out, while cockroaches flourished. Biosphere II became uninhabitable and the £121 million experiment was abandoned after seventeen months.

This kind of failure should temper our belief that there are no limits to our ingenuity. We are not invincible. There are times when our faith in science and technology is both unrealistic and dangerous.

Unrealistic because the world isn't as simplistic as we used to believe. Descartes, the 'father of modern philosophy', famously wrote that he would kick his dog, simply for the pleasure of 'hearing the machine creak'. We now know that there isn't a computer model anywhere that can fully account for the sheer complexity of natural systems. In the real world, recreating them is therefore out of the question.

Dangerous because we cling to the hope that no matter how recklessly we behave, technology will save us. We don't need to change. If we render a species extinct, we can clone a new one. If climate change becomes too severe, we can float giant mirrors in space. And if all else fails, there's always the possibility of colonizing the moon, or if we're really ambitious, discovering a planet that will sustain human life. . .

True, if trends continue, we may well need another planet – or two – but I believe most people favour a different course; one that requires no faith in our ability to cram evolution into a single generation. We may invest billions in space exploration, but of all the problems in the world, not being able to walk on Mars isn't top of most people's list.

When I read Neil Armstrong's claim that 'the important

achievement of Apollo was that it demonstrated that man was not forever *chained* to this planet', it seemed to me to be an extraordinary statement. I wanted to send him a box set of David Attenborough films and ask him if he really views this planet as a place of bondage. After all, what do we need that the natural world does not offer?

We have come to accept that the crisis is real. Now we must learn how to nurture and value what we have; to create a constant economy; one that is a reflection of the real economy, nature's economy.

Sources and Bibliography

'A Blueprint for Survival', *Ecologist*, Vol. 2, No.1, 1972

The Unsettling of America: Culture and Agriculture, Wendell Berry (Sierra Club Books, San Francisco, 1996)

The Millennium Ecosystem Assessment Report, 30 March 2005

Ancient Futures: Learning from Ladakh, Helena Norberg-Hodge (Sierra Club Books, San Francisco, 1992)

The Economics of Climate Change: The Stern Review, Nicholas Stern (CUP, Cambridge, 2007)

Hot Commodities, Jim Rogers (Wiley, London, 2005)

World Disasters Report, International Red Cross, www.ifrc.org

Carbon Down, Profits Up, Climate Group, 2007, www.theclimate-group.org

Open letter to Prime Minister Tony Blair, The Corporate Leaders Group on Climate Change, 2006

WWF's Living Planet Report (www.wwf.org.uk, 2006)

The Trap, James Goldsmith (Macmillan, London, 1993)

Economics in a Full World, Herman E. Daly, *Scientific American*, Vol. 293, Issue 3, September 2005

Affluenza, Oliver James (Vermilion, London, 2007)

Ecological Economics and Sustainable Development: Selected essays of Herman E. Daly (Edward Elgar Publishing, London, 2007)

When Corporations Rule the World, David C. Korten (Kumarian Press, Sterling VA, 2001)

Natural Capitalism: Creating the Next Industrial Revolution, Paul Hawken, Amory B. Lovins and L. Hunter Lovins (Little, Brown, New York, 1999)

What are rainforests worth? And why it makes sense to keep them standing, The Global Canopy Programme: Forest Foresight

Report (www.globalcanopy.org, 4 March 2008)

Power to the People: The Report of Power, An Independent Inquiry into Britain's Democracy, Isobel White (Parliament and Constitution Centre, London, 2006)

Election 2005: Turnout. How many, who and why? (Electoral Commission, London, 2005)

'Audit of Political Engagement 5', Ipsos Mori poll (2008), www.ipsos-mori.com

'Public Finds Much To Support In Conservative's New Green Agenda', Ipsos Mori poll (2007), www.ipsos-mori.com

Happiness: Lessons from a New Science, Richard Layard (Penguin, London, 2005)

Citizen Renaissance, Jules Peck and Robert Phillips, www.citizen-reinaisance.com

The Transition Handbook: From Oil Dependency to Local Resilience, Rob Hopkins (Green Books, London, 2008)

'The Monsanto Files', *Ecologist*, Vol. 28, No. 5, September/October 1998

'GM: Seeds of Change or Fools Gold?', *Ecologist* special issue, November 2008

Genetic Roulette: The Documented Health risks of Genetically Engineered Foods, Jeffrey M.Smith (Yes! Books, Fairfield IA, 2007)

Blueprint for a Green Economy: Submission to the Shadow Cabinet: Quality of Life Policy Group, John Gummer and Zac Goldsmith, September 2007

'Cancer: Are the Experts Lying?', *Ecologist*, Vol. 28, No. 2, March/April 1998

'Synergistic effect of chemical mixtures: Can we rely on traditional toxicology?', Vyvyan Howard, *Ecologist*, Vol. 27, No. 5, September/October 1997

The Big Down: From Genomes to Atoms; Technology converging at a Nano-scale, ETC Group, www.etcgroup.org

'Why the Future Doesn't Need Us', Bill Joy, *Wired*, April 2000

Nanoscience and Nanotechnologies: Opportunities and Uncer-

tainties, Royal Society and Royal Academy of Engineering, 2004

'Secondary Sexual Characteristics and Menses in Young Girls Seen in Office Practise: A Study from the Paediatric Research and Office settings Network', Marcia E. Herman-Gidons et al, *Paediatrics*, Vol. 99, No. 4, April 1997

'Sir Richard Doll: A Questionable Pillar of the Cancer Establishment', Martin Walker, *Ecologist*, Vol. 28, No. 2, March/April 1998

Cancer: Disease of Civilization? An Anthropological and Historical Study, Vilhjalmur Stefansson (Hill & Wang, New York, 1960)

The Mortality from Cancer Throughout the World, Frederick L. Hoffman (Prudential Press, New York, 1915)

The International Assessment of Agricultural Knowledge, Science and Technology for Development, UNESCO, United Nations Food and Agriculture Organization and the World Bank, 2008

An Inconvenient Truth about Food – Neither Secure nor Resilient, Soil Association, November 2008

'An assessment of the total costs of UK agriculture', J. N. Pretty, C. Brett, D. Gee, R. E. Hein, C. F. Mason, J. I. L. Morrison, H. Raven, M. D. Rayment and G. Van der Bijl, *Agricultural Systems*, 2000

Ghost Town Britain (New Economics Foundation, 2002)

Shopped: The Shocking Power of British Supermarkets, Joanna Blythman (Fourth Estate, London, 2004)

Not on the Label: What Really Goes into the Food on your Plate, Felicity Lawrence (Penguin, London, 2004)

The Killing of the Countryside, Graham Harvey (Vintage, London, 1998)

Human health threatened as farm use of lifesaving antibiotics increases again, Richard Young, Soil Association, August 2008

Monocultures of the Mind: Biodiversity, Biotechnology and Scientific Agriculture, Vandana Shiva (Zed Books, London, 1993)

The End of the Line: How Overfishing is Changing the World and

What We Eat, Charles Clover (Ebury Press, London, 2005)

'Rapid worldwide depletion of predatory fish communities', Ransom A. Myers and Boris Worm, *Nature*, Vol. 423, 2003

The Unnatural History of the Sea, Callum Roberts (Gaia Books, London, 2007)

The Low Carbon Economy, The Conservative Party, Policy Green Paper No. 8, January 2009

Automobile: How the Car Changed Life, Ruth Brandon (Macmillan, London, 2002)

Streets for People, Bernard Rudofsky (Doubleday, New York, 1969)

No More School Run: Proposal for a National Yellow Bus Scheme in the UK, Policy Exchange, 2005

'EA's new role must help "minimize flooding heartache"', Countryside Alliance, June 2008

Reforming Energy Subsidies, UNEP, August 2008

The Case Against the Global Economy, and for a turn towards the local, eds Jerry Mander and Edward Goldsmith (Sierra Club Books, San Francisco, 1996)

The Future of Waste Management, DEFRA Select Committee, 2003, www.publications.parliament.uk

'Environmental refugees: An emergent security issue', Norman Myers, Economic forum, Prague, OSCE, May 2005

Millennium Ecosystem Assessment (2005); Liser (2007), www.agassessment.org

Employment and Other Economic Benefits from Advanced Coal Electric Generation with Carbon Capture and Storage, BBC Research and Consulting, February 2009, www.bbcresearch.com

Additional Sources

Carbon Trust: www.carbontrust.co.uk

ClientEarth: www.clientearth.org

Congress for the New Urbanism: www.cnu.org

Ecologist magazine: www.theecologist.org

Energy Saving Trust: www.energysavingtrust.org.uk
Green Alliance: www.green-alliance.org.uk
Hadley Centre: www.metoffice.gov.uk
Insurers' Perspective in Relation to Climate Change:
www.munichre.com
Intergovernmental Panel on Climate Change (IPCC): www.ipcc.ch
OCEANA: Protecting the World's Oceans: www.oceana.org
Oil Change International, Exec Director Stephen Kretzmann:
www.priceofoil.org
Optimum Population Trust (OPT): www.optimumpopulation.org
School Food Matters: www.schoolfoodmatters.com
Soil Association: www.soilassociation.org
Sustainable Development Commission: www.sd-commission.org.uk
World Alliance of Decentralized Energy: www.localpower.org

Note: Sums of money given in US dollars in the source material I have drawn upon have been converted to pounds sterling for consistency, using the exchange rate obtained at the time of typesetting (Spring 2009), of approximately $1.65 to the pound.

Acknowledgements

The purpose of this book is to present ideas that work. We can be sure that they work because the majority of them are already working in various parts of the world. They just aren't working under one umbrella.

On the whole, I have pinched what I regard to be the most important examples in order to lay out a coherent programme. As such, there are more individuals and groups to acknowledge than is possible. However, there are some people whose influence runs throughout the book.

John Gummer MP, with whom I oversaw the Conservative Party's Quality of Life policy review, is a radical pragmatist, and his energy knows no bounds. In the course of our work, we agreed on most issues, though not all, and many of the ideas in this book were uncovered during the course of our work. Jules Peck, who directed the Quality of Life review, helped me develop the first draft of this programme, though it has undergone major changes, and he is not to blame for any of its contents.

Douglas Smith, Harriet Williams and Tony Juniper, all critically reviewed early drafts of the book and made invaluable suggestions. Paul Kingsnorth and, later, Stuart Evers both had a go at editing the chapters, and both improved them considerably. My thanks to Jonathan Pegg, an old friend who took the risk also of acting as my agent, and Toby Mundy and Caroline Knight at Atlantic, who took the risk of publishing it.

Thanks finally to Sheherazade, who read various drafts, and whose comments helped shape the final product.

Index